THE INTERPLANETARY MEDIUM

THE INTERPLANETARY MEDIUM

PART II
OF
SOLAR-TERRESTRIAL PHYSICS/1970

COMPRISING THE PROCEEDINGS OF THE
INTERNATIONAL SYMPOSIUM ON SOLAR-TERRESTRIAL PHYSICS
HELD IN LENINGRAD, U.S.S.R.
12–19 MAY 1970

Sponsored by COSPAR, IAU, IUGG-IAGA, and URSI

E. R. DYER, J. G. ROEDERER, AND A. J. HUNDHAUSEN
Editors

E. R. DYER
General Editor of the Proceedings

SPRINGER-SCIENCE+BUSINESS MEDIA, B.V

ISBN 978-90-277-0211-1   ISBN 978-94-010-3128-8 (eBook)
DOI 10.1007/ 978-94-010-3128-8

TABLE OF CONTENTS

COMPOSITION AND DYNAMICS OF THE SOLAR WIND PLASMA

A. J. HUNDHAUSEN

University of California, Los Alamos Scientific Laboratory, Los Alamos, N.M., U.S.A.

Abstract. This review will concentrate on the interpretation of recent observations pertaining to the chemical composition and dynamical behavior of the solar wind plasma.

The solar wind helium content has been determined in several independent series of observations; the long-term average abundance of helium relative to hydrogen is close to 4.5 % by number. This ratio is lower than that usually given for the solar photosphere, possibly implying a separation of ions with different charge-to-mass ratios in the coronal expansion. Such a separation could result from gravitational settling of heavy coronal ions. A helium-rich plasma has been frequently observed some hours after the passage of a flare-associated interplanetary shock wave; this plasma has been tentatively identified as material ejected from the chromosphere or low corona by the flare. The isotope ^{3}He has been observed by two independent techniques and found to be $\sim 10^3$ times less abundant than ^{4}He. Several ions of ^{16}O have also been identified in the solar wind, with varying total abundance. The observed ionization state of oxygen can be related to the coronal temperature.

Recent observations have established the ranges and average values of the fluid parameters used to describe the solar wind. Attempts to explain observed solar wind temperatures within the framework of fluid models of the coronal expansion require hypotheses of dynamic processes which deposit or redistribute energy in inteplanerary space. Some of the conflicting ideas regarding these processes will be discussed.

The transient behavior of the solar wind has attracted increasing observational and theoretical attention. Numerous observations of interplanetary shock waves indicate that the typical shock wave propagates at ~ 500 km sec^{-1} at 1 AU. In addition to this extreme class of transient phenomena, other long-term patterns of spatial or temporal variation have been found. The common occurrence of such patterns raises questions as to the applicability of the usual spherically symmetric, steady-state models of the coronal expansion in the interpretation of many solar wind observations.

1. Introduction

Comprehensive reviews of solar wind observations and theory were given at the first STP Symposium by Lüst (1967) and Parker (1967). The four years since that symposium have seen the publication of no small number of additional reviews of similar scope; e.g., Dessler (1967), Ness (1967), Wilcox (1968), Hundhausen (1968), Parker (1969), and Holzer and Axford (1970). Rather than adding to this largess, I will attempt here a highly selective discussion of observational and theoretical developments since the 1966 symposium, concentrating on several topics which have led to interesting speculations or quantitative interpretations regarding the physical processes operative in the interplanetary plasma.

First among the topics selected for discussion will be the chemical composition of the solar wind. This subject has often been discussed in near isolation from the general context of coronal and interplanetary dynamics. Although such an approach is a reasonable starting point, the interpretation of many recent observations requires detailed consideration of physical processes occurring in the coronal expansion. Among recent observations pertaining more directly to solar wind dynamics, attention will be focused on the thermal properties of solar wind protons and electrons and the comparison of the observed properties with the predictions of theoretical models.

Dyer (ed.), Solar Terrestrial Physics/1970: Part II, 1–31. All Rights Reserved.

Finally, transient phenomena in the solar wind will be discussed, with concentration on the propagation characteristics of interplanetary shock waves.

2. Chemical Composition of the Solar Wind

2.1. THE AVERAGE RELATIVE ABUNDANCE OF ^{4}He AND ^{1}H

Although the detection of ^{4}He^{++} in the solar wind has been reported by many observers (e.g., Snyder and Neugebauer, 1964; Wolfe and Silva, 1965; Coon, 1966; Lazarus et al., 1966; Neugebauer and Snyder, 1966; Wolfe et al., 1966; Hundhausen et al., 1967a; Ogilvie et al., 1968b; Ogilvie and Wilkerson, 1969; and Robbins et al., 1970), our knowledge of the long-term average interplanetary helium abundance is based on data obtained by three spacecraft: Mariner 2, Vela 3, and Explorer 34.

As an example of these data Figure 1 shows a positive-ion spectrum measured under typical solar wind conditions on Mariner 2. The presence of two flux peaks, the smaller at twice the energy per charge of the larger, was a persistent feature of the observed spectra. These peaks were attributed to ^{1}H$^+$ and ^{4}He^{++} ions travelling at a common mean speed; the helium ions then have twice the energy per charge of the hydrogen

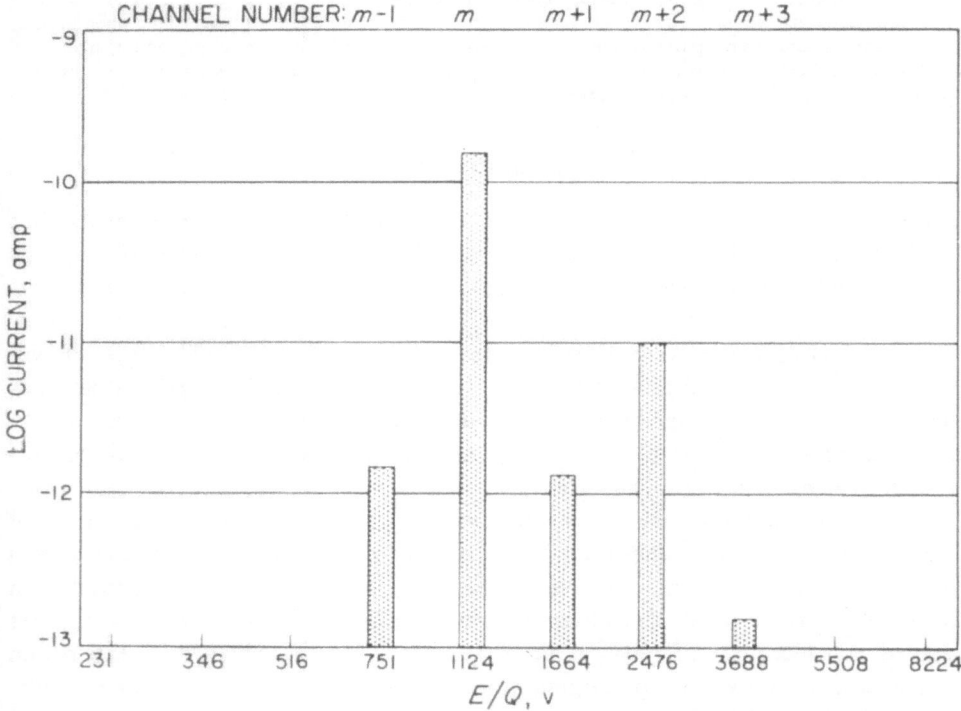

Fig. 1.   A typical solar wind energy-per-charge spectrum observed on the Mariner 2 spacecraft. The primary spectral peak (at channel number m) is attributed to ^{1}H$^+$; the secondary spectral peak (at channel number $m+2$) is attributed to ^{4}He^{++} with the same flow speed as the ^{1}H$^+$ (Neugebauer and Snyder, 1966).

ions, producing an appropriately located secondary spectral peak (see Snyder and Neugebauer, 1963, 1964; and Neugebauer and Snyder, 1966). Note that no measurable flux peak occurs at four times the energy-per-charge of the primary peak, indicating that little $^4He^+$ is present. The relative helium and hydrogen flux or density can thus be determined from data such as that displayed in Figure 1 by comparing the areas under the two spectral peaks. Similar energy-per-charge spectra are obtained by Vela 3 (Hundhausen *et al.*, 1967a; and Robbins *et al.*, 1970) and Explorer 34 (Ogilvie and Wilkerson, 1969). On the latter spacecraft an $E \times B$ velocity selector also permits a mass analysis of the solar wind ions (see Ogilvie *et al.*, 1968a, 1968b). Figure 2 shows typical Explorer 34 solar-wind energy-per-charge spectra at the separate masses corresponding to hydrogen and helium. The conventional attribution of the primary and secondary flux peak to $^-H^+$ and $^4He^{++}$ is confirmed.

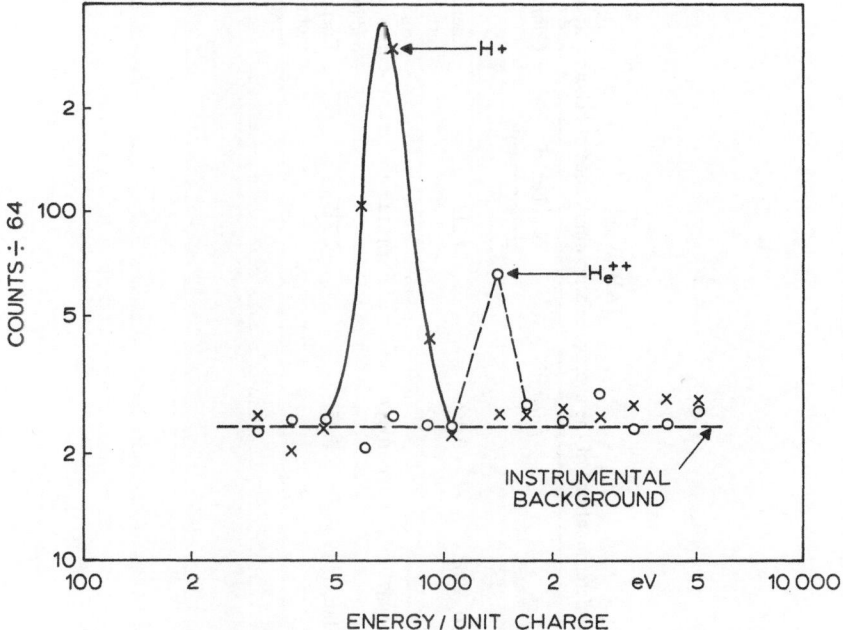

Fig. 2. Typical energy-per-charge spectra at the masses corresponding to $^1H^+$ (crosses) and $^4He^{++}$ (circles) obtained on the Explorer 34 satellite. The conventional attributions of the primary spectral peak to $^1H^+$ and the secondary spectral peak to $^4He^{++}$ are confirmed (Ogilvie *et al.*, 1968b).

Table I summarizes the average helium abundance determined from these three sources. The close agreement between three independent determinations made at separate times argues against both instrumental errors in the measurements and large changes in the average relative abundance of helium and hydrogen between 1962 and 1967. We thus regard the average helium-to-hydrogen density ratio as being well established and near 4.5 %.

TABLE I

Observational determinations of the average relative abundance of helium and hydrogen in the solar wind

Source	Time Period	$\langle n_{He}/n_H \rangle$	Number of measurements	Comments
Mariner 2	Aug. 29 – Dec. 30, 1962	0.046	1 213	Energy-per-charge analysis only; assumptions concerning temperatures; 10% of data included in analysis; sample favors low solar wind speeds
Vela 3	July 1965 – July 1967	0.037	10 314	Energy-per-charge analysis only; 60% of data included in analysis
Explorer 34	May 30, 1967 – Jan. 1, 1968	0.051	2 705	Energy-per-charge and mass analysis; 5% of data included in analysis; sample favors high solar wind density

2.2. COMPARISON OF THE SOLAR AND SOLAR-WIND HELIUM ABUNDANCE

The solar origin of the interplanetary plasma suggests that a relationship (but not necessarily a simple one) will exist between the chemical composition of the outer layers of the Sun and that of the solar wind. The comparison of observed compositions is a natural first step in the examination of this relationship. Unfortunately, the solar abundance of helium can be determined only by indirect means. Three such indirect methods, involving model stellar interiors (Sears, 1964; Demarque and Percy, 1964; Weymann and Sears, 1965; and Morton, 1968), solar cosmic rays (Gaustad, 1964; Biswas and Fichtel, 1964; Lambert, 1967; and Durgaprasad et al., 1968), and the solar neutrino flux (Iben, 1969), yield helium-to-hydrogen abundance ratios (by number) ranging from 0.05–0.09. If these abundance ratios are taken at face value, they indicate that helium is less abundant in interplanetary space than in the Sun. However, the spread in derived solar values probably indicates an uncertainty at least as large as this apparent difference.

2.3. GRAVITATIONAL SETTLING OF HEAVY IONS IN THE CORONA

The comparison of observed interplanetary and solar helium abundances given above suggests that hydrogen may escape from the Sun in the coronal expansion more easily than does helium. A mechanism capable of producing such a separation has been known for some time (see Parker, 1963a, and references therein). In an ionized atmosphere, the tendency for charge separation due to the unequal ratio of gravitational and pressure gradient forces on electrons and ions produces, and is counteracted by, an electric field. If ions of differing charge-to-mass ratios are present, separation of these ions will occur in both static and expanding situations.

Figure 3 shows solutions of the conventional fluid equations (see Parker, 1963b) obtained by Geiss et al. (1970a) for the expansion speeds of $^1H^+$ and $^4He^{++}$ in the presence of such an electric field. The $^4He^{++}$ ions, feeling a larger 'effective gravity' because of the smaller influence of the outward-pointing electric field, expand at a considerably lower speed than the $^1H^+$ ions near the Sun. Collisions between the two ion species, which are also included in this model, lead to nearly equal flow speeds at large heliocentric distances. Figure 4 shows the helium-hydrogen number density ratio in the corona implied by the expansion speeds of Figure 3 and the observed ratio of 0.05 at 1 AU. An appreciable concentration of helium is predicted in the corona, with the density ratio reaching values above 0.2 for $r < 1.2\ R_\odot$. The validity of these solutions near $r = 1$ is questionable; the large temperature gradients in the chromosphere will lead to diffusion of the ions, which is not included in the models of Geiss et al. Despite this limitation, the effect of gravitational settling is explicitly demonstrated by these solutions, and a large concentration of helium in the corona is implied.

The relationship between the *photospheric* and *coronal* helium abundances can be established only by models which are valid in the chromosphere and lower corona, where diffusion terms are important (or even dominant) in the transport equations.

For treatments of these equations see Jokipii (1966), Delache (1965, 1967) and Nakada (1969). Each of these authors predicts an enhancement in the abundance of elements heavier than hydrogen as photospheric material rises through the chromosphere into the corona. The effects of convection, local inhomogeneities, and magnetic fields make the quantitative results of these models uncertain (and explains the decision

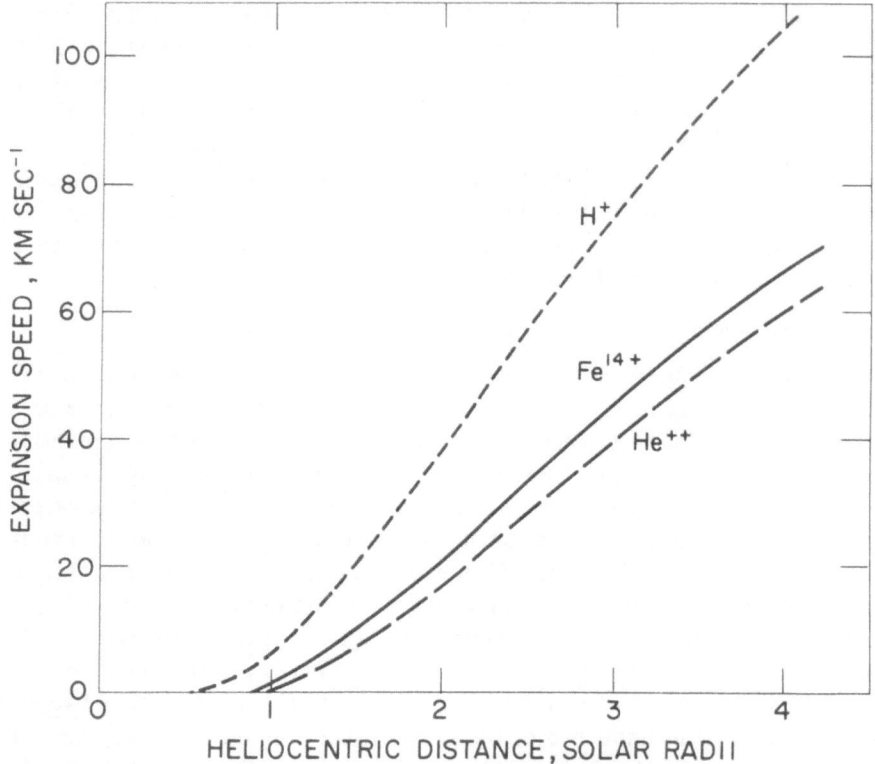

Fig. 3.   The expansion speeds of $^1H^+$, $^4He^{++}$, and $^{56}Fe^{+14}$ as functions of heliocentric distance near the base of the corona. These solutions are for a coronal temperature of 10^6 K. For the $^4He^{++}$ solution, the proton flux at 1 AU is 6.8×10^8 cm^{-2} sec^{-1}; for the $^{56}Fe^{+14}$ solution, the proton flux at 1 AU is 3.4×10^8 cm^{-2} (adapted from Geiss et al., 1970a).

above to relate the coronal helium abundance to that observed in interplanetary space, rather than to a photospheric value). However, an enhanced helium abundance near $r=1$, as in Figure 4, is qualitatively consistent with the predicted effects of diffusion in the chromosphere and lower corona.

2.4. FLUCTUATIONS IN SOLAR-WIND HELIUM ABUNDANCE

Real variations in the solar-wind helium-hydrogen density ratio have been noted by many observers (Coon, 1966; Neugebauer and Snyder, 1966; Hundhausen et al., 1967a; Ogilvie et al., 1968b; Ogilvie and Wilkerson, 1969; and Robbins et al., 1970).

Some basic characteristics of these fluctuations can be inferred from the histograms of observed density ratios based on Vela 3 data (Hundhausen *et al.*, 1967a; and Robbins *et al.*, 1970) and Explorer 34 data (Ogilvie and Wilkerson, 1969): (1) the solar wind helium abundance fluctuates by an order of magnitude; (2) extremely high values are more common than would be expected if the abundances were distributed normally about the mean.

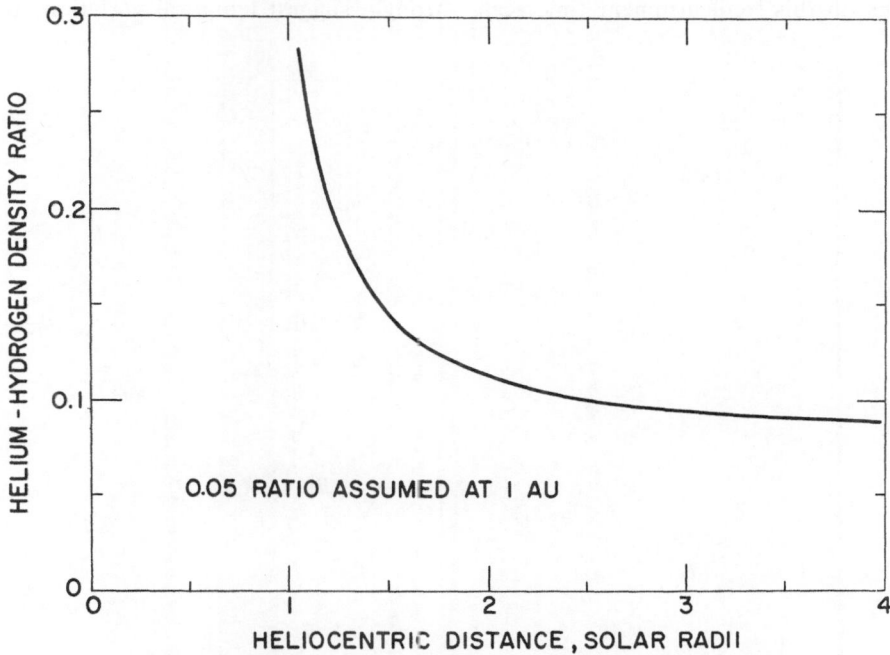

Fig. 4. The abundance (by number) of helium, relative to hydrogen, in the solar corona implied by the ^{4}He^{++} expansion speed curve of Figure 3. Equal ^{4}He^{++} and ^{1}H$^+$ speeds and the observed relative abundance of 0.05 have been assumed at 1 AU.

Further information regarding these variations can be gained by averaging the helium abundance over different time intervals. For example, Figure 5 shows the helium-hydrogen density ratio, observed by Vela 3 satellites between July 19, 1965 to June 28, 1967, averaged over 27-day (dark line) and 10-day (light line) intervals. Averaging over the 27-day solar rotation period removes any inhomogeneities in solar longitude, thus giving the best spatial average over solar wind from the entire corona that is possible with measurements made near the ecliptic plane. These averages show a small trend toward higher values from 1965–1967; this change may be related to the rise in the general level of solar activity. The 27-day averages also show large fluctuations; extreme values are 0.02 and 0.065. Consecutive solar rotations can give widely different helium abundances. These fluctuations and differences imply variations in coronal conditions and a resulting variation in helium abundance on the same spatial scale as the corona itself.

The 10-day averaging period is the approximate time required for transit of a fluid parcel from the base of the corona to 1 AU. All of the material initially within a Sun-centered sphere of radius 1 AU flows out of the sphere in this time and must be replenished by material from lower layers of the solar atmosphere. The 10-day averages of Figure 5 show even larger fluctuations than do the 27-day averages. Fluctuations on this 'replenishment time scale' strongly suggest temporal variations of a

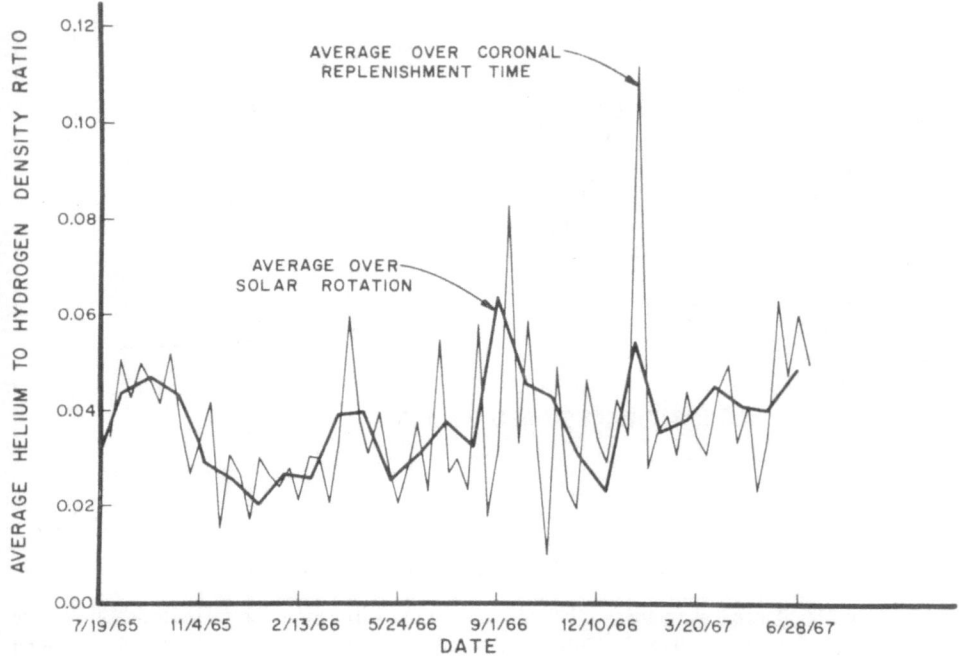

Fig. 5.   The average over twenty-seven-day solar rotation periods and ten-day coronal replenishment periods of the helium-to-hydrogen density ratio observed on Vela 3 satellites. (Robbins *et al.*, 1970.)

similar magnitude in the coronal helium abundance. These coronal variations could be produced either by changes in the relative efficiency with which helium and hydrogen are carried away in the coronal expansion, or by changes in the helium content of the material replenishing the corona. However, 10-day averages could also reflect spatial variations in the coronal helium abundance, produced by local changes in the mechanisms mentioned above. In either case, considerable variation of the coronal chemical composition with position and/or time is implied.

The examination of solar and solar-wind conditions during those 10-day periods when the observed solar-wind helium abundance is high again suggests a relationship between high solar-wind helium content and solar activity. For example, the average helium-hydrogen density ratio for the period March 21–30, 1969 was 0.060 (Figure 5). This concentration of helium was observed in a prominent high-velocity stream which was also present on the previous solar rotation, but with no comparable enhancement of the helium content. The stream was not observed on the following rotation. Two sudden commencements were recorded during these ten days, on March 23 and 25. Interplanetary shock waves associated with both sudden commencements were directly observed by Vela 3. The high-velocity stream and shocks observed at 1 AU were probably produced by a large active region on the Sun (McMath plage region 8207). This region was present on the previous solar rotation, developed into 'the first great active center of this cycle (*HAO Preliminary Report on Solar Activity*)' during the ten days under discussion, and showed little activity during the following solar rotation. The parallel histories of this plage area and the observed interplanetary disturbances strengthen this suggested source identification. The same plage area was independently proposed as the source of the solar wind observed on March 22 and 23 on the basis of an ionization temperature inferred from Vela 3 measurements of oxygen ions (Bame *et al.*, 1968a; and Hundhausen *et al.*, 1968b).

Similar conclusions regarding fluctuations in the coronal chemical composition and the relationship of high interplanetary helium abundance to solar activity can be drawn by examining daily averages of the observed helium concentration. For example, Vela 3 daily averages ranges from 0.003–0.232 (D. E. Robbins and S. J. Bame, private communication). This range is larger than for the 10- or 27-day averages discussed above, and shows that extreme fluctuations (to both low and high values) occur on a time scale of a day or less. Robbins *et al.* (1970) found fourteen days on which the average helium-hydrogen density ratio was greater than 0.08. Thirteen of these occurrences were accompanied within a day by a sudden commencement geomagnetic storm, and thus presumably by an interplanetary shock wave. The shocks associated with ten of these events have been directly observed on Imp 3 (Taylor, 1969) or Vela 3 satellites.

On still shorter time scales, the appearance of plasma with an extremely high helium content five to twelve hours after an interplanetary shock or a geomagnetic sudden commencement has been reported by several observers (Gosling *et al.*, 1967; Bame *et al.*, 1968b; Ogilvie *et al.*, 1968b; Lazarus and Binsack, 1969; Ogilvie and Wilkerson, 1969; and Hirshberg *et al.*, 1970). These observations add to the evidence already given relating high interplanetary helium abundance to solar activity and confirm the more specific connection with flare-associated interplanetary disturbances. As an example of this phenomenon, Figure 6 shows Vela 3A positive-ion energy-per-charge spectra obtained on February 15–16, 1967. An interplanetary shock wave was detected at ≈2345 UT on February 15 both in the Vela 3 plasma data and by the magnetometer on the nearby Explorer 33 satellite (Hirshberg *et al.*, 1970). Both before and shortly after the arrival of the shock, the observed solar-wind helium-hydrogen density ratio was unusually low (0.01–0.02), as indicated by the small secondary flux peak in each of the first two spectra of Figure 6. Nine hours after shock arrival the solar wind speed had increased above the immediate post-shock value, but the helium

content remained low, as indicated by the third spectrum, Figure 6. Shortly there-
after the inbound Vela spacecraft passed through the Earth's bow shock and was no
longer able to observe the undisturbed interplanetary solar wind. At 0916 UT a solar
wind spectrum was again observed, with a drastic change in the secondary flux peak.
This spectrum, the fourth in Figure 6, results from a memory mode of data acquisition,
with widely spaced energy channels which preclude a quantitative determination of the
helium abundance; the relative helium-hydrogen density, however, was clearly
greater than 10%. A sudden worldwide increase of the geomagnetic field was measured

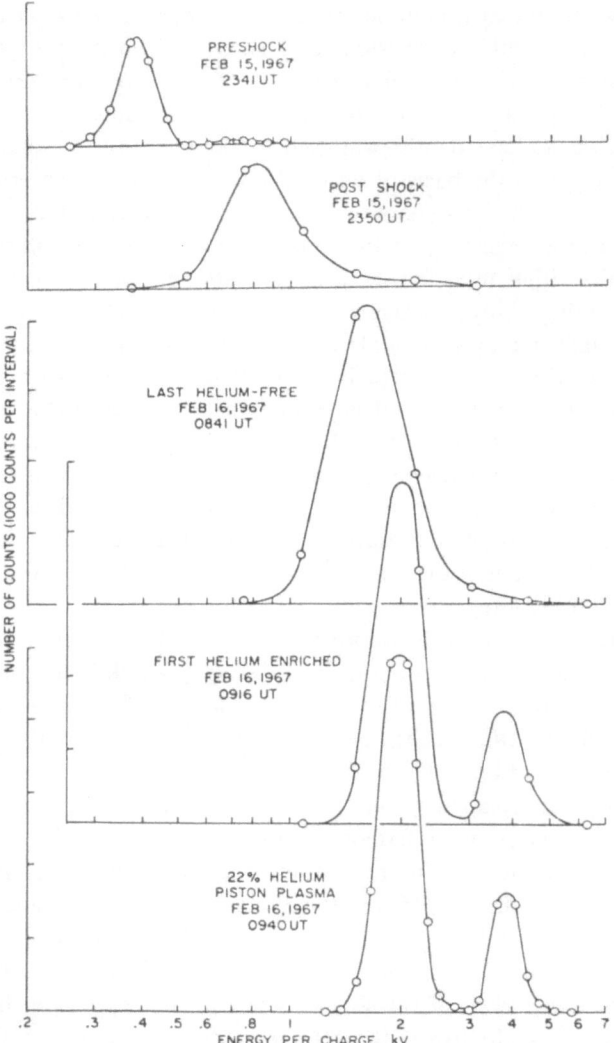

Fig. 6.   Solar-wind energy-per-charge spectra observed by Vela 3A during the solar wind disturbance
of February 15–16, 1967 (Hirshberg *et al.*, 1970).

on ground magnetometers near the time of the helium enrichment at Vela 3A, indicating that a *sudden* change in solar wind conditions had compressed the magnetosphere and left Vela 3 once again exposed to the solar wind. Similar spectra were obtained until 0940 UT when direct data transmission resumed and the fifth spectrum of Figure 6, with normally spaced energy channels, was acquired. A large helium peak, closely resembling that in the memory mode spectra obtained since 0916 UT, was present, from which the helium-hydrogen density ratio is computed to be 0.22. The helium content declined in the following twenty minutes to normal values of 0.05. The helium content of the solar wind is judged to have been above 0.10 for a total of thirty minutes. The interpretation of this helium enrichment is clarified by the simultaneous observations of the interplanetary magnetic field made on Explorer 33. After the abrupt rise in field magnitude associated with passage of the shock, unusually high fields were observed until ≈ 0910 UT, when the magnitude abruptly decreased and the field direction shifted. These changes, essentially coincident with the helium enrichment, can be attributed to the passage of a tangential discontinuity in the interplanetary plasma (see Colburn and Sonett, 1966). Hirshberg *et al.* (1970) have interpreted the simultaneous observation of this discontinuity and the sudden appearance of helium-rich plasma as due to the arrival of the 'driver gas' originally ejected by the flare which produced this interplanetary disturbance. The material observed immediately after the shock but before the tangential discontinuity was originally in the ambient solar wind but had been swept up and compressed by the shock wave before reaching 1 AU. It thus shared the low helium content of the ambient plasma. The flare origin of the material following the tangential discontinuity accounts for its radically different helium content. The region of helium-rich flare ejecta was $\approx 10^6$ km thick at the point of observation.

The observation of interplanetary plasma with a helium-hydrogen density ratio as high as the 0.22 in the example above has interesting implications regarding the coronal helium abundance. Unless a mechanism can be invoked which accelerates $^4\mathrm{He}^{++}$ more efficiently than $^1\mathrm{H}^+$, the coronal helium abundance of the material ejected into the solar wind must have been greater than or equal to that observed at 1 AU. The observation of extreme helium enrichments in the solar wind thus implies a similar or greater enrichment in some coronal region. Note that such values near 20% are consistent with those predicted by the model of Geiss *et al.* (1970a) in the lower corona. The occurrence of a flare near such a coronal region may lead to the ejection of some of the helium-rich material, which is subsequently observed in interplanetary space.

2.5. Observations of other ions

Our discussion of solar wind chemical composition has thus far been based entirely upon observations of $^1\mathrm{H}^+$ and $^4\mathrm{He}^{++}$ ions, shown to be flowing near 1 AU with essentially the same speed. The presence of other ions with mass Mm_p and charge Qq (where m_p is the proton mass, q the electronic charge) would, under the assumption that the flow speed is again equal to that of the $^1\mathrm{H}^+$ ions, result in additional flux peaks in solar wind spectra (such as in Figures 1 and 2) at M/Q times the energy per charge of the primary hydrogen peak. The detection of such ions is difficult because of the rarity in solar material of all other elements and isotopes relative to ^{1}H and ^{4}He, and because the spectral peaks produced by ions with similar values of M/Q will lie close together and can be resolved only if all such peaks are narrow; i.e., if the ion temperatures are sufficiently low. Under the rare combination of favorable circumstances – high solar-wind flux and low ion temperatures – the Vela 3 analyzers have detected the ions ^{4}He$^+$, ^{3}He^{++}, ^{16}O^{+5}, ^{16}O^{+6}, and ^{16}O^{+7}. The presence of ^{3}He in the solar wind has been confirmed by the Apollo 11 and 12 composition experiments.

The Vela 3 positive-ion energy-per-charge data given in Figure 7 (Bame *et al.*, 1968a) shows every ion thus far observed in the solar wind and will serve to organize

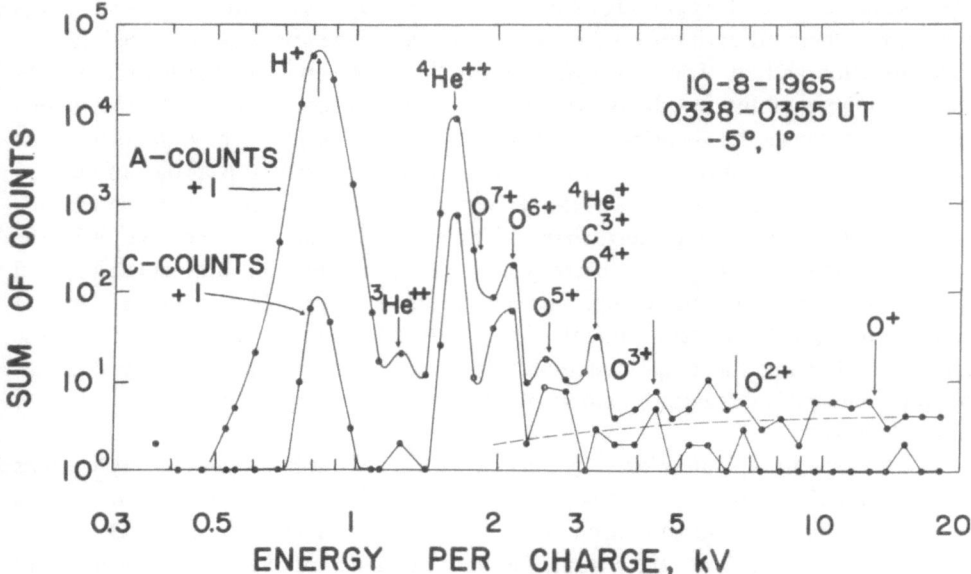

Fig. 7.   Energy-per-charge spectra observed with low pulse thresholds (*A*-counts) and high pulse thresholds (*C*-counts) by Vela 3A. The attribution of various spectral peaks to specific ions is discussed in the text (Bame *et al.*, 1968a).

the present discussion. Two spectra are shown, labelled '*A*-counts' and *C*-counts'. These represent the number of electron multiplier pulses accumulated during the same time intervals, but with two different pulse-size thresholds. A comparison of the two counting levels at a given energy-per-charge channel gives a crude pulse size analysis which will prove useful in identifying ions. The primary spectral peak due to ^{1}H$^+$ appears at ≈ 0.8 kV. The secondary spectral peak produced by ^{4}He^{++} ions occurs, as expected, near 1.6 kV. Other distinct spectral peaks are present; using the M/Q determined by their location and the rough indication of M given by the ratio of A and C counts, the following identifications have been proposed.

(1) The small spectral peak at ≈ 1.2 kV implies the presence of an ion with $M/Q=$

3/2. This can only correspond to $^3He^{++}$. Bame *et al.* (1968a) estimate the abundance of ^3He to be 10^{-3} that of ^4He at the time of this observation. The resolution of the $^3He^{++}$ spectral peak clearly requires that the neighboring $^1H^+$ and $^4He^{++}$ peaks be very narrow, or that the temperature be very low. On another occasion when the prevailing temperature should have made the detection of $^3He^{++}$ possible, the absence of a spectral peak implied an ^3He abundance less than 2×10^{-4} that of ^4He. Buehler *et al.* (1969) and Geiss *et al.* (1970b) have detected ^3He in an aluminum foil exposed to the solar wind during the Apollo 11 and Apollo 12 lunar landings. An abundance of 4×10^{-4} that of ^4He has been determined from the latter foil exposure (Geiss *et al.*, 1970b); this is lower than that from the former. These results confirm the Vela 3 identification of ^3He in the solar wind, as well as the general level of and variability in the abundance of this isotope.

(2) A prominent spectral peak occurs near 2.1 kV, implying the presence of an ion with M/Q near 2.6. A comparison of the A and C spectra indicates that an ion of an element heavier than ^4He is present and eliminates the possibility that some feature in the $^4He^{++}$ distribution has produced has produced the observed spectral peak. The most abundant such elements are ^{16}O, ^{12}C, ^{20}Ne, and ^{14}N, so that the ions $^{16}O^{+6}$, $^{12}C^{+5}$, or $^{20}Ne^{+8}$ are likely candidates. The location of this spectral feature (it is the most often observed peak other than those due to $^1H^+$ and $^4He^{++}$) under varying solar wind conditions is most consistent with an identification as $^{16}O^{+6}$, and Bame *et al.* (1968a) have thus proposed that this ion is the third most abundant in the solar wind. This result is consistent with measured solar abundances of these elements. Note that no spectral peak is observed at the expected position for $^{16}O^{+7}$, while a very small peak is present at the expected position for $^{16}O^{+5}$. Thus most of the solar wind oxygen has been ionized to charge $6q$.

The solar wind abundance of ^{16}O implied by this identification is 1/25 that of ^4He. An hour later a value of 1/66 is obtained. Similar values are obtained at other times, again indicating a variability in the oxygen abundance. The ionization state is generally similar to that implied by Figure 7, with $^{16}O^{+5}$ consistently less abundant than $^{16}O^{+6}$, and $^{16}O^{+6}$ generally more abundant than $^{16}O^{+7}$. The latter ion is, however, occasionally observed to be most abundant. The implications of this ionization state will be mentioned in Subsection 2.6.

(3) Another prominent spectral peak occurs at ≈ 3.2 kV, corresponding to $M/Q=4$. A number of ions, such as $^4He^+$, $^{16}O^{+4}$, or $^{12}C^{+3}$ could produce a feature at this location. However, the ratio of A and C counts at this peak is similar to that at the $^4He^{++}$ peak but not similar to that at the $M/Q \approx 2.6$ peak, tentatively identified as due to the heavier ion, oxygen. This suggests attribution of the $M/Q=4$ peak to $^4He^+$, with the ratio of $^4He^+$ to $^4He^{++}$ densities near 3×10^{-3}. The absence of this spectral feature at other times implies lower ratios and again illustrates the variability of solar wind ionic composition.

2.6. The Ionization State of the Interplanetary Plasma

At the low densities which prevail throughout interplanetary space, the characteristic time scales for atomic processes such as ionization or recombination are extremely long. It would thus be expected (e.g., Brandt and Hodge, 1964; Delache, 1965; Brandt and Hunten, 1966; and Cloutier, 1966) that the ionization state of the expanding solar plasma does not change in interplanetary space, but is entirely determined by initial conditions in the corona. This conclusion can be made more specific by considering the pertinent scale times as functions of position in the corona. The Vela 3 observations discussed above suggest oxygen as an example of interest, and we note from Tucker and Gould (1966) that the most abundant oxygen ions at coronal temperatures will be O^{+6} and O^{+7}. Figure 8 shows τ_6, the time for collisional ionization of

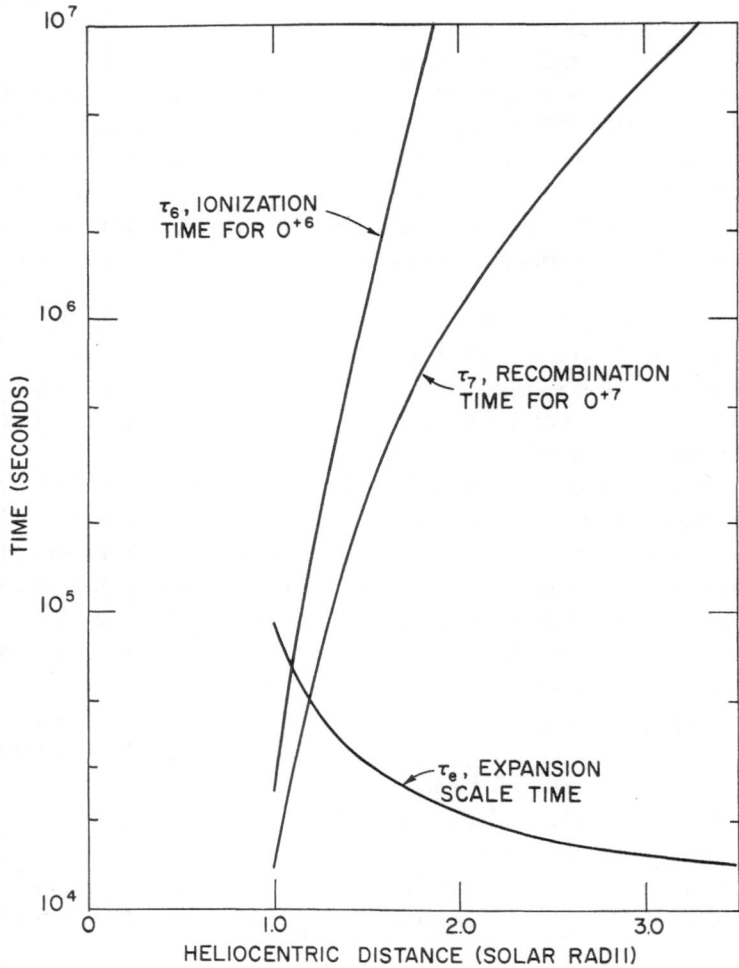

Fig. 8.   Scale times for the ionization of O^{+6}, the recombination of O^{+7}, and the expansion through a density scale height, as a function of heliocentric distance. The density, flow speed, and temperature given by the Whang and Chang solar-wind model have been used (Hundhausen *et al.*, 1968a, b)

O^{+6}, τ_7, the time for radiative (including dielectronic) recombination of O^{+7}, and τ_e, the time for expansion through a density scale height; densities, flow speeds, and temperatures from the coronal expansion model of Whang and Chang (1965) have been used. The scale times τ_6 and τ_7 are much less than τ_e near the base of the corona, indicating that the ionization state of oxygen will adjust to local conditions in this region. However, these scale times increase very rapidly with increasing heliocentric distance (due to their inverse dependence on density, which falls off rapidly as a function of heliocentric distance), becoming equal to the expansion time at $\approx 0.2\ R_\odot$ above the base of the corona and much larger not far above this level. One thus expects the ionization state of the expanding coronal material to remain constant beyond a heliocentric distance of $\approx 1.5\ R_\odot$. Integration of the ionization equations by Hundhausen et al. (1968a, 1968b) and Kozlovsky (1968) confirm this expectation.

Thus the temperature implied by the observed ionization state of a given element in the solar wind can be directly related to the temperature deep in the corona. This relationship opens the interesting possibility of using a measured 'ionization temperature' as a tracer of coronal regions that are sources of the solar wind. For an example of this method, see Hundhausen et al. (1968b).

3. Observations of Solar Wind Properties

Solar wind observations were reviewed by Lüst (1967) at the Belgrade STP symposium; more recent work has been described by Axford (1968) and Hundhausen (1968). For detailed discussions of instrumentation, solar wind data, and the results of solar wind observations, the reader is referred to these reviews. For our present purposes a brief statement of current observational knowledge of solar wind properties will suffice. After a description of the distribution functions of interplanetary protons and electrons, attention will be focused on defining a 'quiet' or 'undisturbed' state of the solar wind. The basic properties of this state will be tabulated for comparison with the predictions of theoretical models.

3.1. DISTRIBUTION FUNCTIONS OF SOLAR WIND PARTICLES

Figure 9 shows a contour mapping of a typical proton velocity distribution function derived from Vela 3 solar wind data (see Hundhausen et al., 1967a, 1967c). The mean velocity of the particles (i.e., the solar wind velocity) is indicated by the small triangle. One of the most interesting features of this observed distribution is related to the non-circular nature of the contours; the random motions of the protons are not isotropic. If a temperature is defined by computing the mean random energy as a function of direction, the value of T_{max} along the direction of elongation (in the lower right quadrant) is about twice the value of T_{min} at 90° to this preferred direction. The average value of the anisotropy magnitude T_{max}/T_{min} observed on Vela 3 is 1.9; the median is 1.6 (Hundhausen et al. 1970a). The Vela observations are 'two-dimensional', giving a projection of the maximum temperature on a plane. Thus the actual anisotropy ratios must be slightly larger.

Such anisotropic distribution functions might be expected because of the conservation of the magnetic moment $v_\perp^2/|\mathbf{B}|$ (where $v_\perp$ is the component of particle velocity normal to the magnetic field $\mathbf{B}$) in the nearly collisionless solar wind flow (e.g., Parker, 1963b). This origin of thermal anisotropies would imply alignment of the direction of $T_{\max}$ with $\mathbf{B}$. The vector $\mathbf{B}_p$ on Figure 9 shows the projection of $\mathbf{B}$, simultaneously measured on the nearby Imp 3 satellite, on the plane of analysis of the Vela 3 detector. The expected alignment is indeed present. Further observations, showing tracking of these two angles, persisting through both gradual and abrupt changes, also show this alignment. A similar conclusion results from a statistical analysis (Hundhausen *et al.*, 1967c). Thus the conservation of $v_\perp^2/|\mathbf{B}|$ is a plausible explanation for the observed proton anisotropies.

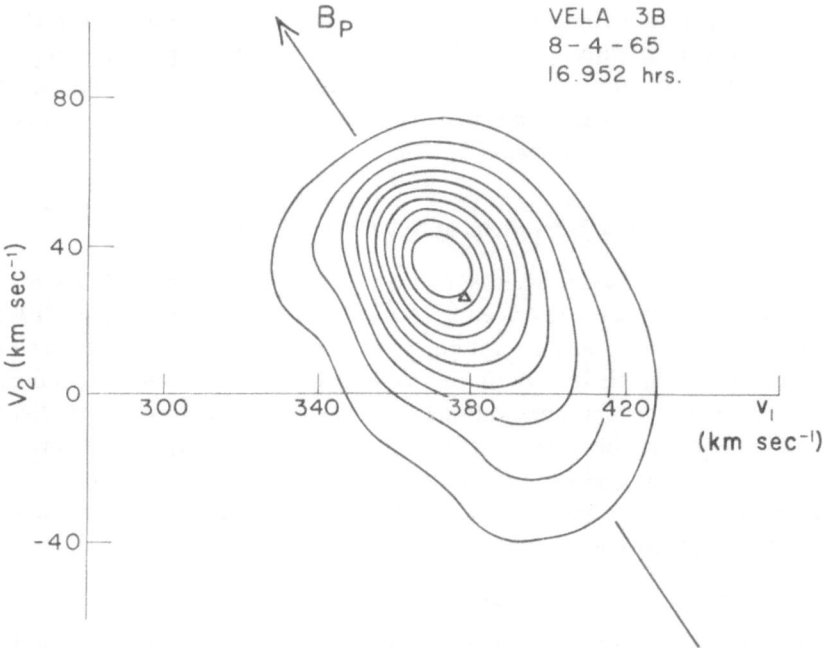

Fig. 9.   A contour mapping of a typical solar-wind proton velocity distribution function observed by Vela 3 satellites. The small triangle at $v_1 = 380$, $v_2 = 20$ is the mean or flow speed. The vector $\mathbf{B}_p$ is the projection on this coordinate plane of the magnetic field observed simultaneously at IMP 3 (Hundhausen *et al.*, 1967c).

In addition to predicting the development of anisotropies in solar wind proton velocity distribution functions, Parker (1963b) pointed out that such anisotropies could lead to instabilities. More specifically, inside of 1 AU, where the interplanetary magnetic field is more nearly radial, the growth of anisotropies with the pressure $P_\parallel$ along the field lines greater than the pressure $P_\perp$ transverse to the field would eventually lead to the firehose instability; outside 1 AU, where the interplanetary magnetic field becomes more nearly azimuthal, the growth of anisotropies with $P_{\parallel} < P_{\perp}$ would lead to the mirror instability. Such instabilities would not only limit the growth of thermal anisotropies and generate waves in the quiet magnetic field, but could also replace Coulomb collisions as a mechanism for maintaining the fluid behavior of the plasma.

The subsequent observation of thermal anisotropies in the solar wind has stimulated considerable interest in such instabilities. The observation of small anisotropies indicates that some limiting mechanism must be operative (Hundhausen et al., 1967a). Of the two instabilities suggested by Parker, the firehose is pertinent to the observed anisotropies, which have the direction corresponding to maximum temperature aligned with **B**. However, the firehose instability will occur only for $\beta_{\parallel} - \beta_{\perp} > 2$, where $\beta_{\parallel} = (8\pi nk T_{\parallel})/B^2$ and $\beta_{\perp} = (8\pi nk T_{\perp})/B^2$, (e.g., Eviatar and Schulz, 1970). The observed anisotropies are generally not large enough to produce this instability (even when the contribution to $\beta_{\parallel}$ and $\beta_{\perp}$ from electrons is included). Numerous other possible instabilities have been suggested, and this topic is one of high current interest.

Figure 10 shows a single slice through a typical solar-wind electron distribution function (derived from Vela 4 data) in the approximate direction of the interplanetary

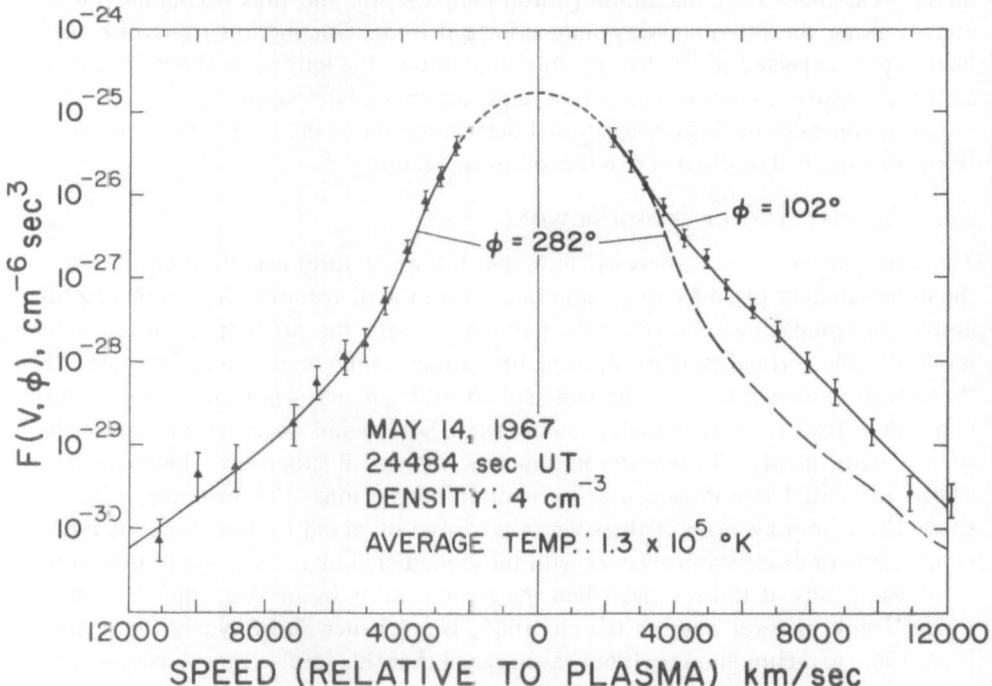

SOLAR WIND ELECTRON DISTRIBUTION FUNCTION

Fig. 10.   A cut through a solar-wind electron velocity distribution function derived from Vela 4B data. In the frame of reference moving with the solar wind speed, the non-Maxwellian tails on the distribution function are asymmetric, implying a transport of energy by heat conduction.

magnetic field. The derived values of the distribution are shown as a function of random speed; i.e., in a frame of reference moving with the plasma. Analysis of such distribution functions leads to the following conclusions (Montgomery *et al.*, 1968):

(1) The electron temperature is higher than the proton temperature by a factor of 3–4 under quiet conditions. The electron temperature generally remains in the range 10^5 to 1.5×10^5 K despite large fluctuations in the proton temperature.

(2) The electron thermal anisotropies are small: T_{max}/T_{min} is generally less than 1.2.

(3) The distribution functions are asymmetric in a frame of reference moving with the plasma. In Figure 10, a line has been drawn through the observed values on the left branch (electrons moving toward the Sun) and reflected about the origin as a dashed line at the right. The observed values on the right branch (electrons moving away from the Sun) fall above this dashed line at random speeds above ≈ 3000 km sec^{-1}. By definition there is no mass transport through any surface in the frame moving with the plasma; the observed asymmetry of the distribution function, however, implies that there *is* an energy transport through such a surface. This energy transport, which is nothing more than the conduction of heat, can be quantitatively determined by computing a third central moment of the observed distribution function. The magnitude of the energy transport varies between 0.005 and 0.02 erg cm^{-2} sec^{-1}, with 10^{-2} cm^{-2} sec^{-1} a typical value. This energy transport generally is in the same direction as that of the maximum proton temperature, and thus by implication (see above) along the interplanetary magnetic field lines. Conduction transverse to the field lines is expected to be strongly inhibited due to the long scale times for electron collisions relative to a cyclotron period in the interplanetary plasma.

The significance of this observational determination of the heat conduction rate in the solar wind will be discussed in the following section.

3.2. THE QUIET STATE OF THE SOLAR WIND

The fluid parameters (i.e., density, flow speed, temperature) usually used to describe the interplanetary plasma can be computed from the distribution functions described above. In comparing such observed parameters with the predictions of theoretical models of the coronal expansion, some limitations of the models must be noted. The theoretical models treat a steady-state spherically symmetric flow of coronal material and would lead to a structureless solar wind. Solar wind observations rarely show such a state; nearly all observed parameters undergo fluctuations which can be ascribed to spatial inhomogeneities or temporal variations. For example, Figure 11 shows three-hour averages of the solar wind speed observed by Mariner 2 in 1962. A comparison of observed properties with those predicted by models would thus appear to be valid only at those times when the solar wind is in an ideal 'quiet' or 'undisturbed' state. Several authors (Coon, 1966; Neugebauer and Snyder, 1966; Strong *et al.*, 1966; and Hundhausen, 1968) have argued that this ideal is most nearly achieved when the solar wind speed is low or, to be more specific, near 320 km sec^{-1}. A quantitative demonstration of this 'quiet' nature of low-speed solar wind has been given, using three-hour averages of the flow speed from Vela 3. Figure 12 shows the root-mean-square averages, in 50 km sec^{-1} flow speed intervals, of the difference between successive three-hour averages of the flow speed observed by Vela 3 (Hundhausen, 1970b), plotted against the flow speed. The variability on this time scale is strongly related to the flow speed, rising from an rms variation of 10 km sec^{-1} at a flow speed of 300 km sec^{-1} to a variation of nearly 40 km sec^{-1} at a flow speed of 500 km sec^{-1}. The low-speed criterion for quiet solar wind will be adopted here, with the warning that even under these conditions the solar wind may not be in the structureless state assumed in constructing the models.

Table II summarizes solar wind properties observed under quiet conditions. Similar typical values were given by Axford (1968) and tabulated by Hundhausen (1968).

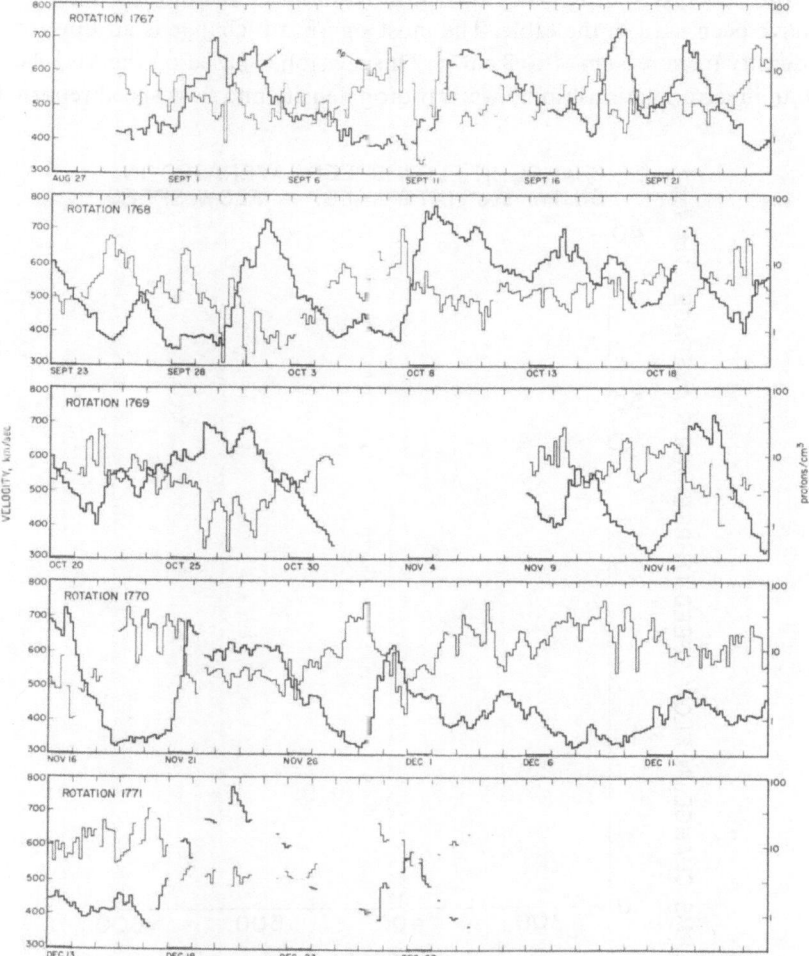

Fig. 11.   Three-hour averages of the solar wind speed (heavier lines) and proton density (lighter lines) observed by Mariner 2 in late 1962 (Neugebauer and Snyder, 1966).

TABLE II

Observed properties in the quiet solar wind

Flow speed	320 km sec^{-1}
Proton or electron density	8 cm^{-3}
Proton temperature	4×10^4 K
Proton thermal anisotropy ratio	2
Electron temperature	1–1.5×10^5 K
Magnetic field	5γ

Minor changes have been made in the light of a recent analysis of Vela 3 observations from July 1965 to July 1967 (Hundhausen *et al.*, 1970a); median values of the solar wind properties observed by Vela 3 when the flow speed fell in the range 300–350 km sec^{-1} have been used in the table. The most significant change is an upward revision of the density from ≈ 5 cm^{-3} to 8 cm^{-3}. This revision is based on the Vela 3 confirmation of an inverse relationship between proton density and flow speed reported earlier

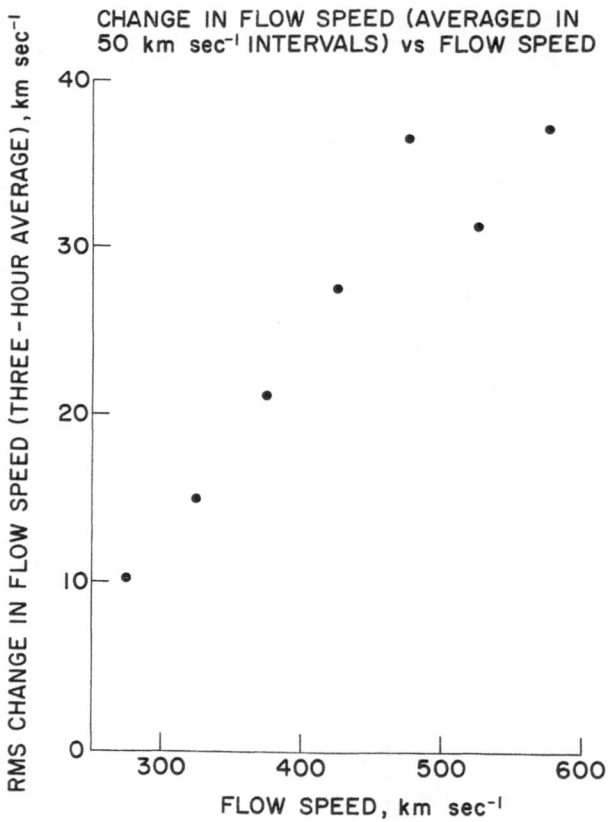

Fig. 12.   The root-mean-square change in consecutive three-hour averages of the flow speed (measured by Vela 3 satellites) as a function of the flow speed (Hundhausen, 1970b).

from Mariner 2 data. A density near 8 cm^{-3} was observed on both spacecraft when the flow speed was near 320 km sec^{-1}.

4. A Comparison of Observations and Theory in the Quiet Solar Wind

As several recent reviews have discussed theoretical models of the formation of the solar wind in the expansion of the corona from basically formal viewpoints (Parker, 1967, 1969; and Holzer and Axford, 1970) and physical viewpoints (Dessler, 1967; and Parker, 1969), there is little reason to repeat this material here. We shall rather concentrate our attention on the discussion of solar-wind dynamical processes which has been stimulated by the comparison of proton temperatures observed at 1 AU with those predicted by theoretical models. Pertinent models are those which integrate an energy equation, assuming some set of energy-source terms as well as the mass and momentum equations. Much of the recent discussion has been based on two classes of models which treat the energy equation in different ways:

(1) One-Fluid Models (Noble and Scarf, 1963; and Whang and Chang, 1965). In one-fluid models the electron and proton temperatures are assumed to be equal, and the energy equation is written assuming that the only energy source beyond the base of the corona is due to heat conduction.

(2) Two-Fluid Models (Sturrock and Hartle, 1966; and Hartle and Sturrock, 1968). In two-fluid models the electron and proton temperatures are not assumed to be equal, and separate energy equations are written for the two species, each with the appropriate conduction source term. The two energy equations are coupled by a conventional exchange term with the energy exchange rate v determined by Coulomb collisions.

Figure 13 shows the temperatures as function of heliocentric distance r obtained in the one-fluid model (dashed line) and two-fluid model (separate solid lines for electrons and protons) with the density and temperature at the base of the corona respectively taken to be 3×10^7 cm^{-3} and 2×10^6 K. As the coronal plasma expands outward from the Sun, the rapidly decreasing density produces a rapid decrease in the energy exchange rate v, and the coupling between the electrons and protons in the two-fluid energy equations becomes weak. The temperatures given by the two-fluid model thus depart from those given by the one-fluid model in a predictable way; the electrons, no longer forced to share internal energy with the protons, remain hot while the protons cool nearly as rapidly as in an adiabatic expansion. Predicted temperatures at 1 AU are 2.1×10^5 K from the one-fluid model, $T_e = 3.5 \times 10^5$ K and $T_p = 4.4 \times 10^3$ K from the two-fluid model. The observed values at 1 AU, $T_e = 1.5 \times 10^5$ K and $T_p = 4 \times 10^4$ K, are also shown on the figure. The observation that $T_e > T_p$ is in agreement with the prediction of the two-fluid model. However, the predicted ratio T_e/T_p is far too high, largely because the predicted proton temperature is an order of magnitude lower than the observed value.

The discrepancy between the proton temperature predicted by the two-fluid model and that actually observed at 1 AU has been interpreted as evidence that an additional 'non-thermal' heat source term should be added to the proton energy equation (Sturrock and Hartle, 1966; Strong *et al.*, 1966; Hundhausen *et al.*, 1967b; and Barnes, 1968, 1969). Such an energy source $S(r)$ could serve to raise both the proton temperature and the flow speed predicted at 1 AU. Since the two-fluid model (with the coronal conditions specified above) also predicts too low a flow speed (250 km sec^{-1}) at 1 AU, both of the changes produced by the added energy term $S(r)$ would improve

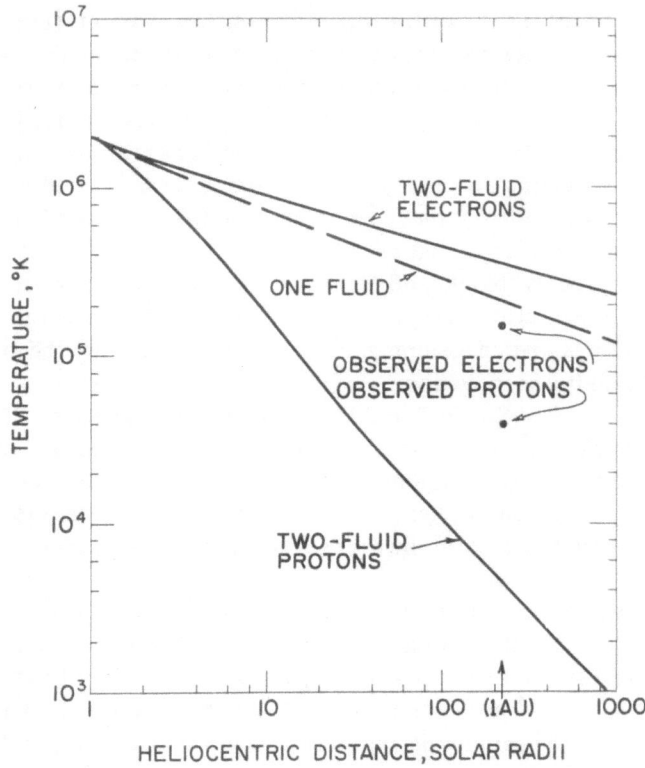

Fig. 13.   Particle temperatures as functions of heliocentric distance predicted by the two-fluid model (solid lines) and one-fluid model (broken line) for a coronal temperature of 2×10^6 K. The proton and electron temperatures observed at 1 AU under quiet conditions are indicated.

the agreement between predictions and observations. For a quantitative discussion of the incorporation of non-thermal heating terms into the two-fluid model, see Hartle and Barnes (1970). A specific physical mechanism, the collisionless damping of hydromagnetic waves emitted from the Sun, for interplanetary heating of the protons has been proposed by Barnes (1968, 1969).

The electron temperature $T_e = 3.5 \times 10^5$ K predicted at 1 AU by the two-fluid model is appreciably higher than the observed value of 1.5×10^5 K. This predicted electron temperature implies a very large transport of energy by heat conduction, and results in a large discrepancy when compared with observed energy fluxes (Hundhausen, 1969b). Whereas the observations indicate that only 4% of the total solar-wind energy flux is due to heat conduction, while 92% of the total is carried in the bulk flow, the two-fluid model gives 55% of the total flux in heat conduction, with only 37% of the total in the bulk flow. In essence, the high electron temperature obtained in the two-fluid model gives a large flux of energy in the form of heat conduction because the conductivity is proportional to $T^{5/2}$. This energy does not become available to the bulk flow of the plasma, and a low flow speed results. In fact, the *total energy flux* predicted by the two-fluid model for the coronal conditions assumed by Hartle and Sturrock is more than twice that actually observed.

This comparison of energy fluxes raises questions regarding the use of an additional source term to raise the flow speed and proton temperature predicted by the two-fluid model. Even though such a source can improve the agreement with observed values of these parameters, it cannot appreciably reduce the heat conduction flux, and can only make the disparity in total energy flux larger. This objection suggests consideration of other mechanisms for raising the predicted proton temperature. For example, several authors (Hundhausen *et al.*, 1967c; Montgomery *et al.*, 1968; Hundhausen, 1968; and Nishida, 1969) have advocated an enhanced rate of energy exchange between the hot electrons and cooler protons. The one-fluid model is, of course, the limiting case for a large exchange rate ν, and does, in fact achieve a partition of the total energy flux much more nearly in agreement with the observations. For the coronal density and temperature assumed above (and used by Hartle and Sturrock in the two-fluid model), the one-fluid model gives 10% of the total energy flow in heat conduction. The predicted flow speed of 335 km sec^{-1} is in much better agreement with the observed value than the 250 km sec^{-1} predicted by the two-fluid model. It should, of course, be recalled that the observations give $T_e/T_p \approx 4$, so that use of an energy exchange rate ν so large that a one-fluid solution results is not justified; the one-fluid model is mentioned here only as a limiting case. The quantitative incorporation of enhanced energy exchange rates into the two-fluid model would be an interesting test of the enhanced coupling mechanism.

5. Time-Dependent Phenomena in the Solar Wind

The comparison of solar wind observations and theories given in Section 4 dealt exclusively with steady-state phenomena. It has been shown (Subsection 3.2) however, that observations rarely reveal steady conditions in the solar wind and that special care must be taken in selecting observations suitable for comparison with theoretical models that assume a time-independent, spherically symmetric coronal expansion. Under the varying conditions generally observed in the solar wind, other dynamical processes specifically associated with temporal variations or spatial gradients may be important. Both the observational and theoretical work in this area has concentrated on two extreme classes of phenomena. The first of these involves the solar wind variations related to localized, long-lived regions of solar activity; i.e., recurrent solar wind streams and related magnetic sectors in interplanetary space. Observations of these spatial structures have been reviewed by Wilcox (1968); for recent theoretical discussions see Carovillano and Siscoe (1969) and Siscoe and Finley (1969a, 1969b). We will concentrate attention here on the second class of time-dependent phenomena, which involves the solar wind variations related to short-lived solar activity; i.e., interplanetary shock waves produced by solar flares. Some final comments will be made concerning the general variability of the solar wind.

5.1. Observations of flare-associated interplanetary shock waves

The occurrence of geomagnetic storms subsequent to some solar flares has long been attributed to the arrival at the Earth of material ejected by the flare. The existence of an ambient interplanetary plasma implies that a shock will propagate ahead of the flare ejecta (Gold, 1955). The first observation interpreted as an interplanetary shock was made on the Mariner 2 spacecraft in 1962 (Sonett *et al.*, 1964, 1966); numerous additional shock observations have been reported in the past few years. The reasonable agreement between the Rankine-Hugoniot relations and the observed changes in solar wind properties (Sonett *et al.*, 1966; Gosling *et al.*, 1968; Ogilvie and Burlaga, 1969; and Chao, 1970) confirms the interpretation of the observations as shocks, and indicates the presence of a strong, collisionless dissipation mechanism in the interplanetary plasma. We will be concerned here with dynamical processes on a larger scale, relating to the propagation of shock waves in the solar wind. The observations will first be summarized and typical characteristics of an interplanetary shock will be deduced. Theoretical models of shock propagation in the solar wind will then be described and used to estimate the magnitude of energy release by a flare.

Figure 14 shows Vela 4B measurements of the solar wind speed, proton density, and proton and electron temperatures (two values of each temperature appear because of the presence of thermal anisotropies) from a seven hour period on June 5, 1967. The abrupt increases in all four parameters at 1915 UT signaled the passage of an interplanetary shock wave which will serve as a typical example. This shock was also observed on the Explorer 34 satellite (Ogilvie *et al.*, 1968b). Figure 14 also shows a crossing of the standing bow shock wave in front of the earth at 1820 UT and a brief sampling of the region of altered flow (the magnetosheath) behind this bow shock. A comparison of the changes in particle temperatures at the interplanetary and bow shock directly indicates that the former is relatively weak.

The observation of pre- and post-shock solar wind properties permits computation of the shock propagation speed. Use of either the Vela 4 or Explorer 34 observed values for the June 5 shock gives $V = 510$ km sec^{-1}. Viewed in the frame of reference moving with the shock front, the upstream plasma flows into the shock at 110 km sec^{-1}. As the speed of sound in the pre-shock region can be computed to be 58 km sec^{-1}, the sonic Mach number of the shock is 1.9. Any of several solar flares that occurred a few days before the observation of the June 5 shock at 1 AU could have produced this solar wind disturbance (see Hundhausen, 1970a). Ogilvie *et al.* (1968b) have favored association of the shock with a flare observed at 0800 UT on June 3. This would give a transit time from the Sun of 59.5 hours and a mean propagation speed of 700 km sec^{-1}, appreciably larger than the 510 km sec^{-1} propagation speed inferred from the observations at 1 AU. This difference may indicate a deceleration of the shock in interplanetary space.

Table III summarizes some characteristics of the interplanetary shocks reported in the literature. Although plasma observations are necessary to compute most of the shock parameters of interest here, the observations made with the Imp 3 magneto-meter (Taylor, 1969) are included, since they represent the largest available sample of

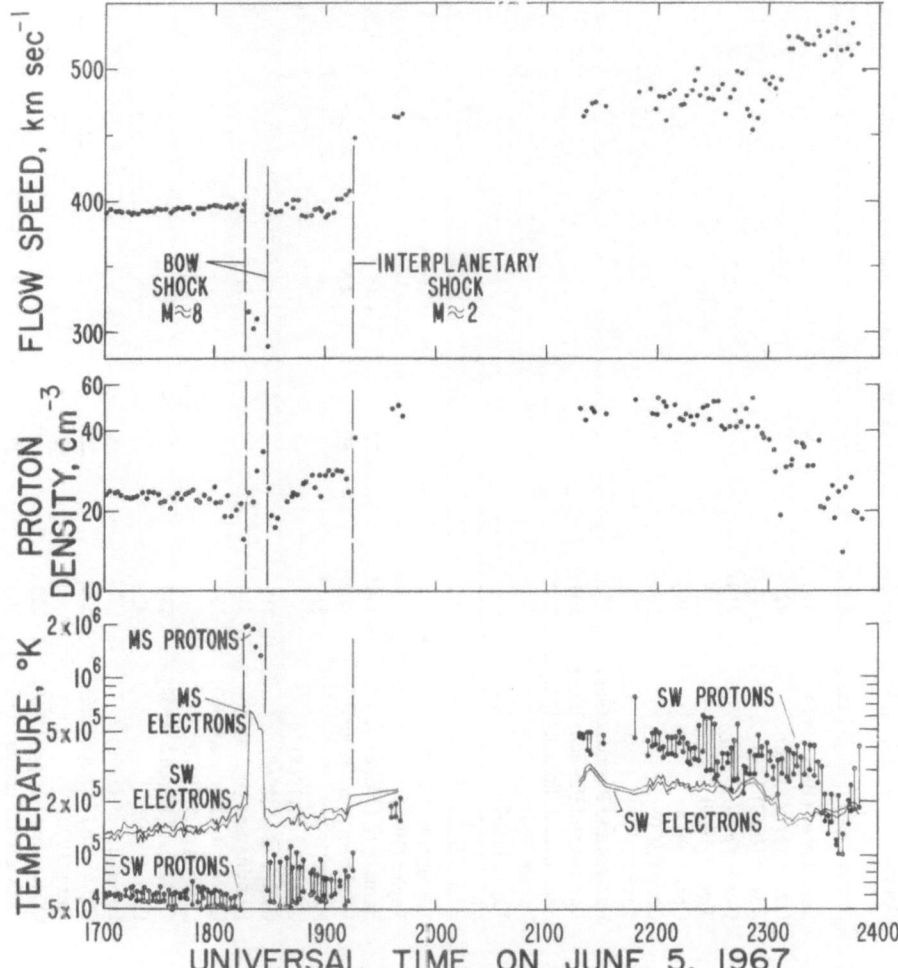

Fig. 14. Solar wind properties (flow speed, proton number density, and proton and electron tempe-rature) observed by Vela 4B on June 5, 1967. The abrupt changes at 1915 UT signal the passage of an interplanetary shock wave. A brief excursion behind the earth's bow shock occurred between 1820 and 1930 UT (Hundhausen, 1970a). In the temperature plot (bottom) MS = magnetosheath, SW = solar wind.

TABLE III

Properties of directly observed interplanetary shock waves

Date	Source	Shock Speed (km sec⁻¹)	Angle between shock normal and radial direction	Sonic mach number	Alfvén Mach number	Transit time (hours)	Mean speed (km sec⁻¹)
Oct. 7, 1962	Sonett et al. (1964, 1966)	510					
Oct. 5, 1965	Gosling et al. (1968)	420	9°	3.0		17.2 or 71*	2500 or 590*
Oct. 7, 1965	Taylor (1969)					49.5	840
Jan. 20, 1966	Gosling et al. (1968)	410		2.3		27.3 or 63.5	1670 or 650
	Taylor (1969)					63.5	650
March 22, 1966	Chao (1970)	480	33°		5.5		
March 23, 1966	Chao (1970)	560	49°		9.2		
July 8, 1966	Taylor (1969)					44.6	930
July 9, 1966	Lazarus and Binsack (1969)	750					
July 10, 1966	Lazarus and Binsack (1969)	500					
July 15, 1966	Lazarus and Binsack (1969)	830					
	Taylor (1969)					102	400
Aug. 29, 1966	Taylor (1969)					67.2	620
	Chao (1970)	470	15°		4.8		
Sept. 3, 1966	Taylor (1969)					39.5	1050
Jan. 6, 1967	Taylor (1969)					55.7	740
Jan. 7, 1967	Taylor (1969)					45.6	910
Jan. 13, 1967	Bame et al. (1968b)	~430*	~70°	~4*		58	720
Feb. 15, 1967	Hirshberg et al. (1970)	480*	60°	4.5*		53.5	660
May 1, 1967	Hones (1970)	510*		3*			
May 30, 1967	Ogilvie et al. (1968)	510				56.2	730
June 5, 1967	Ogilvie et al. (1968)	510				59.5	700
	Hundhausen (1970a)	350		1.5		?	?
June 25, 1967	Ogilvie and Burlaga (1969)						
	Lazarus et al. (1970)						
June 26, 1967	Ogilvie and Burlaga (1969)	480	8°		3.1		
	Chao (1970)	420	33°		8.4		
Aug. 11, 1967	Lazarus et al. (1970)		85°		15		
Aug. 29, 1967	Ogilvie and Burlaga (1969)	500	22°		1.5		
	Chao (1970)	340	47°		1.0		
Sept. 13, 1967	Ogilvie and Burlaga (1969)	420	6°				
Sept. 19, 1967	Ogilvie and Burlaga (1969)	500	27°		1.7		
Jan. 11, 1968	Ogilvie and Burlaga (1969)	520	30°		11.3		
Feb. 26, 1969	Bonetti et al. (1970)	600		2.0			
	Hundhausen et al. (1970c)	570				?	?

flare associations. The columns in the table give: (1) the date of the shock observation, (2) the source of observational information, (3) the shock propagation speed inferred either from the mass continuity equation or from the delay between two spatially separated observations, (4) the angle between the observed normal to the shock front, generally determined from measurement of the change in magnetic field, and the radial direction (for those observations with no angle given, the shock normal was assumed to be radial in the computation of the propagation speed), (5) the sonic Mach number of the shock, (6) the Alfvén Mach number, based on the magnetic field component along the shock normal, (7) the time required for propagation to 1 AU implied by a flare association, and (8) the mean propagation speed corresponding to this time. An asterisk following any entry signifies a value computed by this author from parameters given in the original source or suggested as an alternate value.

The twenty-seven shock observations summarized in Table III form a reasonable statistical sample from which some average characteristics can be deduced. The mean shock propagation speed is 500 km sec^{-1}, or about 100 km sec^{-1} higher than the mean solar wind speed. The shock normal deviates from the radial direction by an average angle of 30° (omitting the extreme case of August 11, 1967). The assumption of a radial shock normal in some of the computations of the shock speed thus leads to an average overestimate of $\approx 15\%$; the value obtained for moderate failures of this assumption is essentially the radial component of the true shock speed. The sonic Mach number is generally near 3, indicating that the typical interplanetary shock is of intermediate strength. The average transit time from the Sun implied by flare associations is 55 hr; this gives a mean propagation speed inside of 1 AU of 730 km sec^{-1}. As this latter value is greater than the average propagation speed at 1 AU, a small deceleration of the shock during transit from the Sun appears to be usual. However, the radial component of the 1 AU shock speed in several cases with large deviations from radial propagation (e.g., January 13 and February 15, 1967) is comparable to the mean speed, indicating that not all shocks are decelerated (see Hirshberg *et al.*, 1970).

5.2. THEORETICAL MODELS OF SHOCK PROPAGATION IN THE SOLAR WIND

The theoretical treatment of propagating solar wind disturbances requires integration of the time-dependent equations for the conservation of mass, momentum, and energy. Even in the one-fluid formulation that has been employed in all models to date the presence of a second independent variable so complicates the integrations that numerous additional simplifications must be made. The flow is generally taken to be spherically symmetric and adiabatic.

Analytical solutions of the time-dependent equations can be obtained by using so-called 'similarity' theory, if solar gravity and the flow of the ambient solar wind are neglected; see Parker (1961, 1963b), Simon and Axford (1966), and Lee and Balwanz (1968). Lee and Chen (1968) have included the magnetic field in the equations, while Korobeinikov (1969) has obtained solutions in which the ambient solar wind speed is not neglected. Numerical solutions of the time-dependent equations, including solar gravity and the flow of the ambient solar wind, have been obtained by Hundhausen and Gentry (1969a, 1969b). Both techniques illustrate the existence of two limiting classes of solutions: (1) 'blast waves', in which the initial energy release into the solar wind disturbance occurs over a time short compared to that required for the disturbance to propagate to an observer; (2) 'driven disturbances', in which the energy release occurs over a non-negligible fraction of the propagation time. In the former class a unique relationship exists between the total energy in the disturbance and the properties of the shock at large heliocentric distances.

5.3. COMPARISON OF OBSERVATIONS AND THEORETICAL MODELS

In comparing the theoretical models briefly described above with observed interplanetary shock waves, the idealized nature of the theory should not be forgotten. Numerous simplifying assumptions (e.g., the equality of electron and proton temperatures, the neglect of magnetic forces, heat conduction, or any dissipation mechanism away from the shock) are common to both the similarity and numerical models and limit their applicability to the interplanetary plasma.

The unique dependence of blast wave properties on the total energy W predicted by both the similarity and numerical models has been used to estimate W from observations made at the shock (Dryer and Jones, 1968; Hundhausen and Gentry, 1969a; and Korobeinikov, 1969). For an average solar-wind density of 8 protons cm^{-3} at 1 AU, the average transit time $\langle T \rangle = 55$ hr deduced above implies an energy $\langle W \rangle = 5 \times 10^{32}$ ergs from similarity theory or $\langle W \rangle = 3 \times 10^{31}$ ergs from the numerical integrations. Similar values result from the average shock speed $\langle V \rangle = 500$ km sec^{-1}. The difference between the energy estimates based on similarity and numerical methods is due largely to the neglect of the ambient solar wind speed in the former. For driven disturbances the energy $\langle W \rangle$ implied by the same $\langle T \rangle$ or $\langle V \rangle$ would be several times larger. Such model-dependent estimates of the energy in flare-associated solar-wind disturbances were until recently the only sources of such information.

Integration of the excess of the energy flux above the pre-shock or ambient level throughout a flare-associated solar wind disturbance gives the energy W added to the solar wind by the flare. For observations made at a single position at heliocentric distance r, the integration can be approximated by assuming that the observed solar wind properties are averages over an area A covered by the disturbance, and integrating over time. Assuming that the disturbance at 1 AU subtends a solid angle of π viewed from the Sun, Hundhausen et al. (1970b) have used Vela 3 and Vela 4 observations to estimate W for nineteen solar wind disturbances. The resulting values of W range from 5×10^{30} ergs to 2×10^{32} ergs; $\langle W \rangle = 5 \times 10^{31}$ ergs. Disturbances resembling both blast waves and driven disturbances were observed. For those solar wind disturbances whose post-shock flow most resembles blast waves, $\langle W \rangle = 1.4 \times 10^{31}$ ergs, in good agreement with the prediction based on numerical solutions.

5.4. GENERAL VARIATIONS IN THE STATE OF THE SOLAR WIND

The time-dependent interplanetary phenomena mentioned above, namely recurrent solar wind streams and flare-associated disturbances, have been extreme categories selected because of their relative simplicity, which facilitates both physical understanding and theoretical treatment. Many variations in solar wind properties are observed which do not fit into these two categories but which probably involve both temporal and spatial variations. Such variability in solar wind conditions will certainly have some effect on basic solar wind properties. For example, Burlaga and Ogilvie (1970) have reported the existence of 'hot spots', regions i.e., wherein the proton temperature is higher than normal. The hot spots exist at large positive velocity gradients, and have been shown (Burlaga *et al.*, 1970) to be qualitatively consistent with adiabatic heating of the ions by compression. This, of course, represents a conversion of energy from the bulk motion of an 'overtaking' solar wind stream to internal energy of the compressed 'overtaken' plasma. Theoretical treatments of such phenomena involve considerable computational difficulties and have received little attention. Siscoe (1969) has considered the effect of fluctuations on solar wind properties by integrating the conservation equations over spherical surfaces. Specific applications of this technique to solar wind models require additional information on the pertinent dynamical processes, which must be deduced from observations or assumed. In view of the observational evidence indicating the continual presence of fluctuations in the solar wind, extension of theoretical models to include such effects would appear to be a fruitful field for future research.

Acknowledgements

The author wishes to thank Drs. D. W. Forslund, R. A. Gentry, J. Hirshberg, and M. D. Montgomery for discussions of material relevant to this review, and Drs. S. J. Bame, M. D. Montgomery, and D. E. Robbins for permission to use unpublished data. This work was performed under the auspices of the United States Atomic Energy Commission.

References

Axford, W. I.: 1968, *Space Sci. Rev.* **8**, 331.
Bame, S. J., Hundhausen, A. J., Asbridge, J. R., and Strong, I. B.: 1968a, *Phys. Rev. Letters* **20**, 393.
Bame, S. J., Asbridge, J. R., Hundhausen, A. J., and Strong, I. B.: 1968b, *J. Geophys. Res.* **73**, 5761.
Barnes, A.: 1968, *Astrophys. J.* **154**, 751.
Barnes, A.: 1969, *Astrophys. J.* **155**, 311.
Biswas, S. and Fichtel, C. E.: 1964 *Astrophys. J.* **134** 941.
Bonetti, A., Moreno, G., Candidi, M., Egidi, A., Formisano, V., and Pizzella, G.: 1970, in *Intercorrelated Satellite Observations Related to Solar Events* (ed. by V. Manno and D. E. Page), D. Reidel Publ. Co., Dordrecht, Holland, p. 436.
Brandt, J. C. and Hodge, P. W.: 1964, *Solar System Astrophysics*, McGraw-Hill, New York.
Brandt, J. C. and Hunten, D. M.: 1966, *Planetary Space Sci.* **14**, 95.
Buehler, F., Eberhardt, P., Geiss, J., and Meister, J.: 1969, *Science* **166**, 1502.
Burlaga, L. F. and Ogilvie, K. W.: 1970, *Astrophys. J.* **159**, 659.
Burlaga, L. F., Ogilvie, K. W., Fairfield, D. H., Montgomery, M. D., and Bame, S. J.: 1970, NASA-Goddard Space Flight Center preprint.
Carovillano, R. L. and Siscoe, G. L.: 1969, *Solar Phys.* **8**, 401.
Chao, J.-K.: 1970, MIT Center for Space Research preprint CSR-TR-70-3.
Cloutier, P.: 1966, *Planetary Space Sci.* **14**, 809.
Colburn, D. S. and Sonett, C. P.: 1966, *Space Sci. Rev.* **5**, 439.

Coon, J. H.: 1966, in *Radiation Trapped in the Earth's Magnetic Field* (ed. by B. M. McCormac), D. Reidel, Publ. Co., Dordrecht, Holland, p. 231.

Delache, P.: 1965, *Compt. Rend. Acad. Sci.* **261**, 643.

Delache, P.: 1967, *Ann. Astrophys.* **30**, 827.

Demarque, P. R. and Percy, J. R.: 1964, *Astrophys. J.* **140**, 541.

Dessler, A. J.: 1967, *Rev. Geophys.* **5**, 1.

Dryer, M. and Jones, D. L.: 1968, *J. Geophys. Res.* **73**, 4875.

Durgaprasad, N., Fichtel, C. E., Guss, D. E., and Reames, D. V.: 1968, *Astrophys. J.* **154**, 307.

Eviatar, A. and Schultz, M.: 1970, *Planetary Space Sci.* **18**, 321.

Gaustad, J. E.: 1964, *Astrophys. J.* **139**, 406.

Geiss, J., Hirt, P., and Lentwyler, H.: 1970a, *Solar Phys.* **12**, 458.

Geiss, J., Eberhardt, P., Singer, P., Buehler, F., and Meister, J.: 1970b, to appear in *Apollo 12 Preliminary Science Report*.

Gold, T.: 1955, in *Gas Dynamics of Cosmic Clouds* (ed. by H. C. van de Hulst and J. M. Burgers), North-Holland Publ. Co., Amsterdam, p. 103.

Gosling, J. T., Asbridge, J. R., Bame, S. J., Hundhausen, A. J., and Strong, I. B.: 1967, *J. Geophys. Res.* **72**, 1813.

Gosling, J. T., Asbridge, J. R., Bame, S. J., Hundhausen, A. J., and Strong, I. B.: 1968, *J. Geophys. Res.* **73**, 43.

Hartle, R. E. and Sturrock, P. A.: 1968, *Astrophys. J.* **151**, 1155.

Hartle, R. E. and Barnes, A.: 1970, NASA-Goddard Space Flight Center preprint X-621-70-114.

Hirshberg, J., Alksne, A., Colburn, D. S., Bame, S. J., and Hundhausen, A. J.: 1970, *J. Geophys. Res.* **75**, 1.

Holzer, T. E. and Axford, W. I.: 1970, *Ann. Rev. Astron. Astrophys.* **8**, 31.

Hones, E. W., Jr.: 1970, in *Intercorrelated Satellite Observations Related to Solar Events* (ed. by V. Manno and D. E. Page), D. Reidel Publ. Co., Dordrecht, Holland, p. 299.

Hundhausen, A. J., Asbridge, J. R., Bame, S. J., Gilbert, H. E., and Strong, I. B.: 1967a, *J. Geophys. Res.* **72**, 87.

Hundhausen, A. J., Asbridge, J. R., Bame, S. J., and Strong, I. B.: 1967b, *J. Geophys. Res.* **72**, 1979.

Hundhausen, A. J., Bame, S. J., and Ness, N. F.: 1967c, *J. Geophys. Res.* **72**, 5265.

Hundhausen, A. J.: 1968, *Space Sci. Rev.* **8**, 690.

Hundhausen, A. J., Gilbert, H. E., and Bame, S. J.: 1968a, *Astrophys. J.* **152**, L3.

Hundhausen, A. J., Gilbert, H. E., and Bame, S. J.: 1968b, *J. Geophys. Res.* **73**, 5485.

Hundhausen, A. J.: 1969a, *J. Geophys. Res.* **74**, 3740.

Hundhausen, A. J.: 1969b, *J. Geophys. Res.* **74**, 5810.

Hundhausen, A. J. and Gentry, R. A.: 1969a, *J. Geophys. Res.* **74**, 2908.

Hundhausen, A. J. and Gentry, R. A.: 1969b, *J. Geophys. Res.* **74**, 6229.

Hundhausen, A. J.: 1970a, in *Particles and Fields in the Magnetosphere* (ed. by B. M. McCormac), D. Reidel Publ. Co., Dordrecht, Holland, p. 79.

Hundhausen, A. J.: 1970b, *Ann. Geophys.* **26**, 427.

Hundhausen, A. J., Bame, S. J., Asbridge, J. R., and Sydoriak, S. J.: 1970a, Los Alamos Scientific Laboratory preprint LA-DC-11329, *J. Geophys. Res.* **75**, 4643.

Hundhausen, A. J., Bame, S. J., and Montgomery, M. D.: 1970b, Los Alamos Scientific Laboratory preprint LA-DC-11203, *J. Geophys. Res.* **75**, 4631.

Hundhausen, A. J., Bame, S. J., and Montgomery, M. D.: 1970c, in *Intercorrelated Satellite Observations Related to Solar Events* (ed. by V. Manno and D. E. Page), D. Reidel Publ. Co., Dordrecht, Holland, p. 567.

Iben, Icko, Jr.: 1969, *Ann. Phys.* **54**, 164.

Jokipii, J. R.: 1966, in *The Solar Wind* (ed. by R. J. Mackin and M. Neugebauer), Pergamon Press, New York, p. 215.

Korobeinikov, V. P.: 1969, *Solar Phys.* **7**, 463.

Kozlovsky, B.-Z.: 1968, *Solar Phys.* **5**, 410.

Lambert, D. L.: 1967, *Nature* **215**, 43.

Lazarus, A. J., Bridge, H. S., and Davis, J.: 1966, *J. Geophys. Res.* **71**, 3787.

Lazarus, A. J. and Binsack, J. H.: 1969, *Annals of the IQSY* (A. Stickland, ed.), MIT Press **3**, 378.

Lazarus, A. J., Ogilvie, K. W., and Burlaga, L. F.: 1970, MIT Center for Space Research preprint CSR-P-70-36, submitted to *Solar Phys.* **13**, 232.

Lee, T. S. and Balwanz, W. W.: 1968, *Solar Phys.* 4, 240.
Lee, T. S. and Chen, T.: 1968, *Planetary Space Sci.* 16, 1483.
Lüst, R.: 1967, in *Solar-Terrestrial Physics* (ed. by J. W. King and W. S. Newman), Academic Press, London, p. 1.
Montgomery, M. D., Bame, S. J., and Hundhausen, A. J.: 1968, *J. Geophys. Res.* 73, 4999.
Morton, D. C.: 1968, *Astrophys. J.* 151, 285.
Nakada, N. P.: 1969, *Solar Phys.* 7, 302.
Neugebauer, M. and Snyder, C. W.: 1966, *J. Geophys. Res.* 71, 4469.
Ness, N. F.: 1967, *Ann. Rev. Astron. Astrophys.* 6, 79.
Nishida, A.: 1969, *J. Geophys. Res.* 74, 5155.
Noble, L. M. and Scarf, F. L.: 1963, *Astrophys. J.* 138, 1169.
Ogilvie, K. W., McIlwraith, N., and Wilkerson, T. D.: 1968a, *Rev. Sci. Instr.* 39, 441.
Ogilvie, K. W., Burlaga, L. F., and Wilkerson, T. D.: 1968b, *J. Geophys. Res.* 73, 6809.
Ogilvie, K. W. and Burlaga, L. F.: 1969, *Solar Phys.* 8, 422.
Ogilvie, K. W. and Wilkerson, T. D.: 1969, *Solar Phys.* 8, 435.
Parker, E. N.: 1961, *Astrophys. J.* 133, 1014.
Parker, E. N.: 1963a, in *The Solar Corona* (ed. by J. W. Evans), Academic Press, New York, p. 11.
Parker, E. N.: 1963b, *Interplanetary Dynamical Processes*, Interscience, New York.
Parker, E. N.: 1967, in *Solar-Terrestrial Physics* (ed. by J. W. King and W. S. Newman), Academic Press, London, p. 45.
Parker, E. N.: 1969, *Space Sci. Rev.* 9, 325.
Robbins, D. E., Hundhausen, A. J., and Bame, S. J.: 1970, *J. Geophys. Res.* 75, 1178.
Sears, R. L.: 1964, *Astrophys. J.* 140, 477.
Simon, M. and Axford, W. I.: 1966, *Planetary Space Sci.* 14, 901.
Siscoe, G. L.: 1970, *Cosmic Electrodyn.* 1, 51.
Siscoe, G. L. and Finley, L. T.: 1969a, *Solar Phys.* 9, 452.
Siscoe, G. L. and Finley, L. T.: 1969b, University of California, Los Angeles, Dept. of Meteorology, preprint.
Snyder, C. W., Neugebauer, M., and Rao, U. R.: 1963, *J. Geophys. Res.* 68, 6361.
Snyder, C. W. and Neugebauer, M.: 1964, *Space Res.* 4, 89.
Sonett, C. P., Colburn, D. S., Davis, L., Jr., Smith, E. J., and Coleman, P. J., Jr.: 1964, *Phys. Rev. Letters* 13, 153.
Sonett, C. P., Colburn, D. S., and Briggs, B. R.: 1966 in *The Solar Wind* (ed. by R. W. Mackin and Briggs, B. R.: 1966 in *The Solar Wind* (ed. by R. W. Mackin and M. Neugebauer), Pergamon Press, New York, p. 165.
Strong, I. B., Asbridge, J. R., Bame, S. J., Heckman, H. H., and Hundhausen, A. J.: 1966, *Phys. Rev. Letters* 16, 631.
Sturrock, P. A. and Hartle, R. E.: 1966, *Phys. Rev. Letters* 16, 628.
Taylor, H. E.: 1969, *Solar Phys.* 6, 320.
Tucker, W. H. and Gould, R. J.: 1966, *Astrophys. J.* 144, 244.
Weymann, R. and Sears, R. L.: 1965, *Astrophys. J.* 142, 174.
Whang, Y. C. and Chang, C. C.: 1965, *J. Geophys. Res.* 70, 5793.
Wilcox, J. M.: 1968, *Space Sci. Rev.* 8, 258.
Wolfe, J. H. and Silva, R. W.: 1965, *J. Geophys. Res.* 70, 3575.
Wolfe, J. H., Silva, R. W., McKibben, D. D., and Mason, R. H.: 1966, *J. Geophys. Res.* 71, 3329.

THE CONFIGURATION OF THE INTERPLANETARY
MAGNETIC FIELD

LEVERETT DAVIS, Jr.

California Institute of Technology, Calif., U.S.A.

Abstract. The idealized basic structure of the interplanetary magnetic field is the familiar spiral wound on a cone whose axis is the solar rotation axis. Variations in the radial velocity of the solar wind produce large scale variations in pitch; slight distortions are also produced by the non-radial component of the solar wind velocity. The high velocity streams in the solar wind seem to be more significant than magnetic polarity alternation in the sector structure of the interplanetary medium. Superposed on the ideal spiral are a variety of smaller structures. Outwardly propagating non-sinusoidal Alfvén waves with scale lengths of order 10^5–10^7 km are common; their characteristics and relation to the high velocity solar wind streams are examined in some detail. When sharp crested, they are recognized as rotational discontinuities. Other, less easily identified, waves are present part of the time. Tangential discontinuities and other convected structures have been identified, as have interplanetary shock waves.

1. Introduction

This review of the structure of the interplanetary magnetic field starts with very simple, very familiar ideas and proceeds to features only recently discovered and still subject to debate. The first topic is the large scale structure, the features whose scale is a substantial fraction of their distance from the Sun. Next, intermediate structures are considered, that is, those features that are very much larger than the proton gyro-radius but very small compared to their distance from the Sun. The solar wind plasma must be examined as well as the magnetic fields because for many features the plasma motions appear to be the basic cause while the magnetic field is partly an easily measurable clue to the plasma behavior and partly a means by which fluid behavior is produced in the plasma. In fact, a major point will be the importance of the high velocity solar wind streams in any study of the magnetic and plasma structure of interplanetary space.

2. Large Scale Structure

The large scale structure may be reviewed briefly. Basically, it is the spiral proposed by Parker (1958, 1963) in his early discussions of the solar wind. It is produced by the solar wind moving radially outward, accelerated near the Sun but of nearly constant velocity throughout most of the solar system. The magnetic field lines remain rooted in the Sun but their outer ends are swept outward, each element of fluid sliding outward along the line of force because of the very high electrical conductivity. The rotation of the Sun carries the roots of the lines around and produces the familiar spirals.

These spirals are wound on the circular cones, shown in Figure 1, whose axis is the Sun's axis of rotation, Ω. Thus the most suitable coordinates to use are solar polar coordinates, the polar axis being Ω. In later figures, we shall use the right-handed, orthogonal *RTN* system, a variant of the familiar notation in which *R* is in the radial direction, *T*, the tangential direction, more often denoted by φ, and *N*, the normal

Dyer (ed.), Solar Terrestrial Physics/1970: Part II, 32–48. All Rights Reserved.
Copyright © 1972 by D. Reidel Publishing Company, Dordrecht-Holland.

direction, opposite to the usual θ-direction and nearly northward except near the poles. In the solar equatorial plane, which makes an angle of 7.º25 with the ecliptic, the cone becomes a plane and the spiral becomes the classic Archimedean spiral in the region where the solar wind velocity is independent of radius.

Since the solar wind velocity varies from day to day, the pitch of the classic spiral varies substantially over distances of the order of $\frac{1}{2}$ AU radially or 30° in solar longitude. Thus the actual spiral becomes quite irregular. This is not usually regarded as of great importance because locally the structure, smoothed to remove intermediate scale structures, is still a spiral of the locally appropriate pitch and the entire large

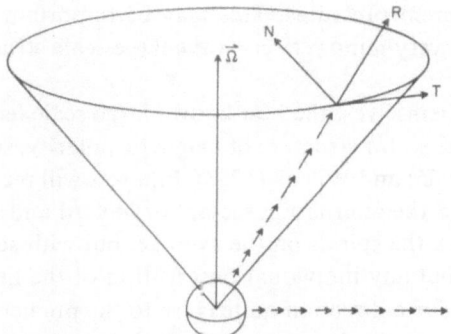

Fig. 1.   The cone on which the magnetic field spirals are wound and the *RTN* coordinate system. The Sun's axis of rotation is Ω and the arrows indicate the solar wind velocity along two possible cones.

scale structure is built out of segments of the ideal spirals. As will be seen below, the dynamical phenomena associated with the fitting together of these segments produce very important features of the intermediate scale structure.

More careful analysis reveals a number of minor modifications of this simple model. They are of little importance for the general large scale structure, but are quite important when considering the angular momentum loss of the Sun or the observed azimuthal motions of the plasma. The model based on purely radial flow assumes that the gas stops its azimuthal motion as soon as it leaves the Sun. If one allows it to continue as required for conservation of angular momentum, it would have an azimuthal velocity of only about 10 m/sec at the orbit of Earth. This is completely negligible compared to the 300–700 km/sec radial velocity. In the model with no azimuthal motion, there are no magnetic forces to perturb the motion; but when angular momentum is conserved, the forces exerted by the magnetic field increase the azimuthal velocity at the orbit of Earth to about 1 km/sec in the direction of the Sun's rotation (Weber and Davis, 1967; and Modisette, 1967). If anisotropic pressure and viscous forces are added to the magnetic and inertial forces, this figure can be increased to about 10 km/sec. Meyer and Pfirsch report at this meeting on one model of this kind. Weber and Davis (1970) have presented another model. It appears that with a reasonable range of choice for the values of the parameters, a wide range of azimuthal velocities, including negative values, is predicted. This is probably appropriate since the observations, which are very difficult to do with precision, seem to show (Hundhausen, 1968) as large a range in azimuthal velocity as do the theories.

Very near to the Sun, within a few solar radii, the magnetic structure is complicated and is grossly affected by the complex sources below and in the photosphere. The plasma flow in much of the lower corona is likely to be dominated by magnetic structures. The location and nature of the high velocity solar wind streams are probably determined in this region, perhaps in part by magnetic structures (Davis, 1966) and in part by the energy supply to the corona. Farther out from the Sun there may be some deflection from radial flow in both the azimuthal and polar directions if adjacent streams have different pressures. This may be important for some phenomena, but should have only a very minor effect on the large scale structure of the interplanetary field.

Next to the spirals themselves, the best known large scale feature of the interplanetary magnetic field is the sector structure of magnetic polarity, so beautifully developed by Ness and Wilcox (1965) and Wilcox (1968). Figure 2 will recall its familiar features. It shows the spirals and the alternating sectors of inward and outward directed magnetic fields which follow the spirals on the average, but with substantial local fluctuations. It is not likely that any individual observation of the field, or even an average over three hours, will give a direction quite close to the predicted spiral. Nevertheless, the basic pattern is the spiral with superposed wiggles.

The polarity in the sectors is determined by the sign of the large scale average of the radial component of the photospheric magnetic field (Wilcox and Howard, 1968). Small scale fluctuations in the photospheric field, as for example in sunspots, more often produce only arches in the corona and make no contribution to the interplanetary magnetic field. Coronal structure may be important in determining the strength and polarity of the field lines that escape and the character of the variable solar wind plasma flow (Davis, 1966). Thus the strength and sense of the magnetic field in the spirals is determined by the slowly evolving sources at the solar surface and by the way these sources connect through the corona to the solar wind region.

In the example shown in Figure 2, there are four sectors, two inward and two outward, for each rotation of the Sun. Sometimes these sectors remain stable for several solar rotations. Sometimes they evolve rapidly (Coleman et al., 1966). Sometimes there are only two sectors per solar rotation. This was found by Mariner 2 in 1962 before the sector concept had been invented and by Mariner 5 in 1967. A great deal of information on this is summarized in Figure 3, which shows the variations in the patterns as the solar cycle progresses. Each horizontal strip summarizes one solar rotation.

An essential feature of the sector concept is that when it is most regular it organizes not only magnetic polarity but also patterns of plasma flow, of the recurring solar cosmic rays, and of geomagnetic disturbance (Wilcox and Ness, 1965). Let us start our review of this by recalling the Mariner 2 observations by Snyder and Neugebauer (1966). They found a series of high velocity streams coming out from the Sun, 4 or

5 per solar rotation. They found that these were long lived and persisted from one solar rotation to the next. And, in collaboration with Rao (Snyder *et al.*, 1963), they found that the high velocity streams tended to be associated with high values of K_p. It was noted very early that one of these streams had one polarity and most of the others the opposite polarity. The Mariner 5 data show a number of high velocity streams for each solar rotation during the summer of 1967, but they are not identifiable from one solar rotation to the next (Bridge *et al.*, 1970). Mostly the polarity is negative and the small sector of positive polarity does not seem to be identifiable with any particular stream.

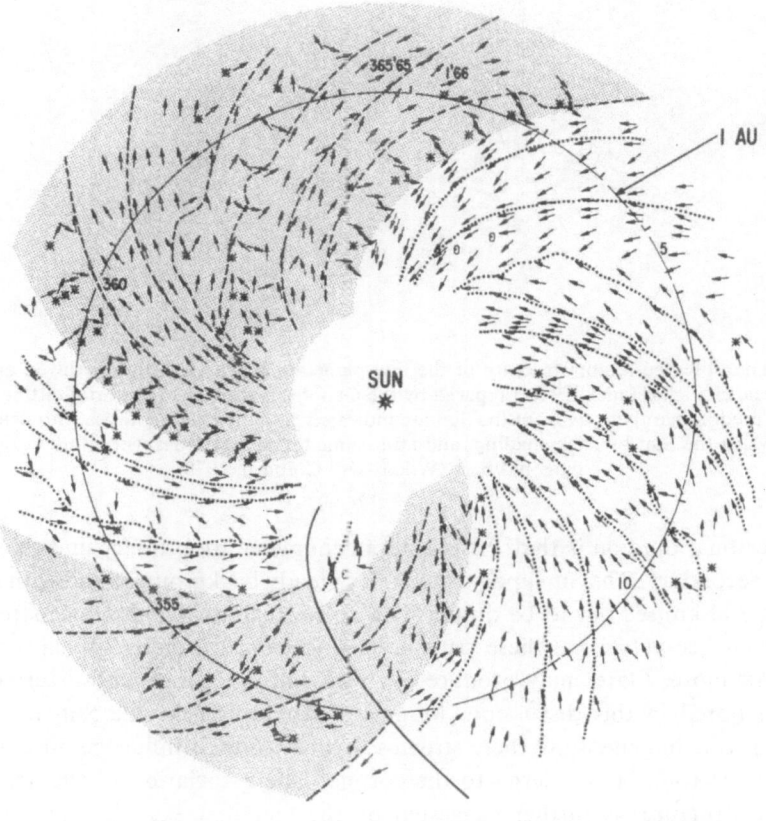

PIONEER 6 ECLIPTIC MAGNETIC FIELD
SOLAR ROTATION 1502
352/65-14/66
DEC. 18, 1965 - JAN. 14, 1966

Fig. 2. An example of sector structure, shown by shading, and the interplanetary magnetic field, shown by arrows, deduced by Schatten *et al.* (1968) by extrapolation from observations at 1 AU.

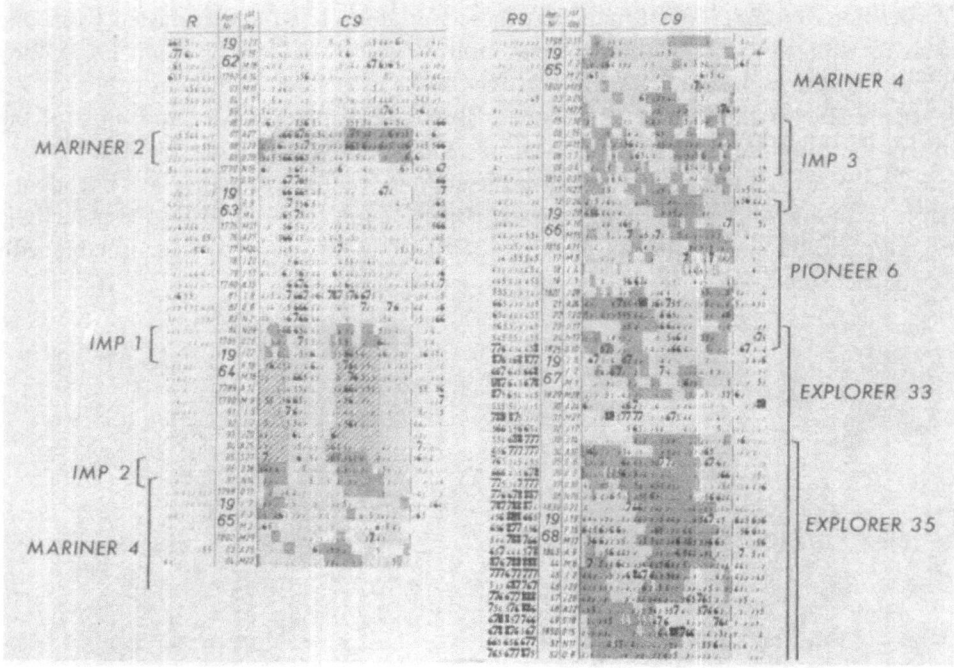

Fig. 3.   The observed sector structure of the interplanetary magnetic field, overlayed on the daily geomagnetic character index *C9* as prepared by the Geophysikalisches Institut of Göttingen. Sectors with field predominantly away from the Sun are indicated by light shading, those with field predominantly toward the Sun by dark shading, and an assumed quasi-stationary structure during 1964 by diagonal bars (Wilcox and Colburn, 1970).

My personal opinion is that the plasma phenomena are primary and the magnetic patterns secondary. The Sun appears to emit a steady background wind with a velocity of about 300 km/sec but to be dotted with active regions that produce streams with 500–700 km/sec velocities, these streams lasting from a number of days to several months or more. Flares may produce vigorous but more transient disturbances that will be ignored in this discussion. Magnetic structures near the Sun may play an important role in generating these streams because of their influence on both the gas flow and the transfer of energy to the corona. The association of the streams with magnetic structures is further suggested by the fact that each stream usually, and perhaps always, has only a single magnetic polarity. Adjacent streams may have opposite or the same polarity, possibly even in a statistically random way. The classical IMP 1 observations showed alternating polarity, but earlier and later examples are more nearly random.

3. Intermediate Scale Magnetic Features

To introduce the discussion of intermediate scale magnetic features, consider the typical data shown in Figures 4. Each part shows the Mariner 5 magnetometer data for one hour on Day 166 of 1967 when very near 1 AU. Figure 4a shows very quiet data, Figure 4b somewhat disturbed data, and Figure 4c quite disturbed data. Note that at all times there are small fluctuations at frequencies that are higher than can be resolved at the sampling rate used; these are mainly real and not instrumental noise. As the frequency becomes lower, the amplitude of the fluctuations typically increases. Since the field is swept past the spacecraft by the wind with a velocity, during this period, of about 400 km/sec, much faster than any relevant wave velocities, these temporal variations in the components are easily translated into a three-dimensional spatial structure of the field lines. Note that there are a number of discontinuities but that many changes are reasonably gradual. Note too that B_A, the scalar magnitude of the field strength, is much more nearly constant than are the components. Thus the spatial structure is one in which there are many changes in the directions of the lines of force but much less change in field strength.

In the parts of Figure 4 each individual data point is shown. In Figure 5 the data have been averaged over short intervals, the averages for the entire day are shown. On this different scale the structures appear very different. There are small changes in the absolute value and substantial changes, some of which look like rough waves, in the components.

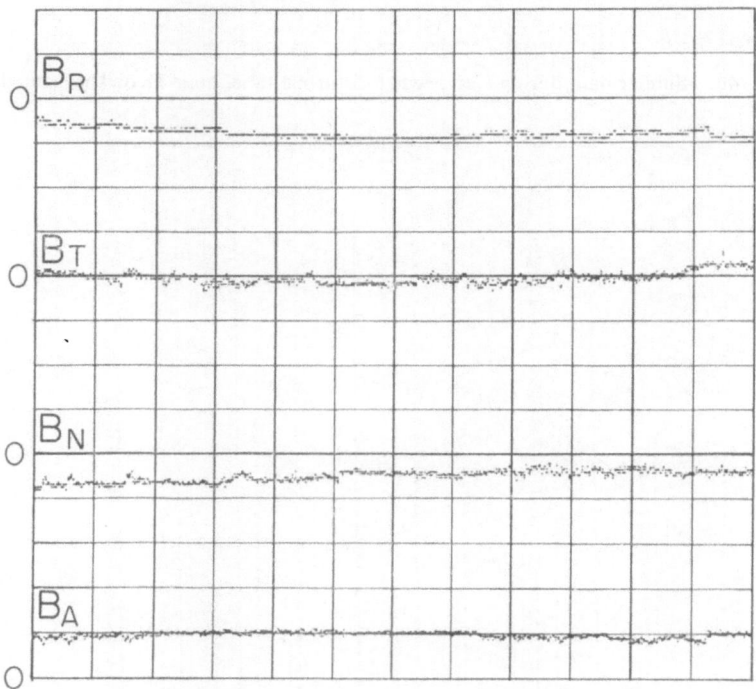

Fig. 4a.   The Mariner 5 magnetometer data during a quiet time, hour 22 on day 166 of 1967. The R, T, and N components in the coordinate system of Figure 1 and the absolute value of the field strength are shown, there being three independent vector observations in each 12.6 sec interval. The distance between horizontal grid lines is 5γ; between vertical grid lines it is 5 min.

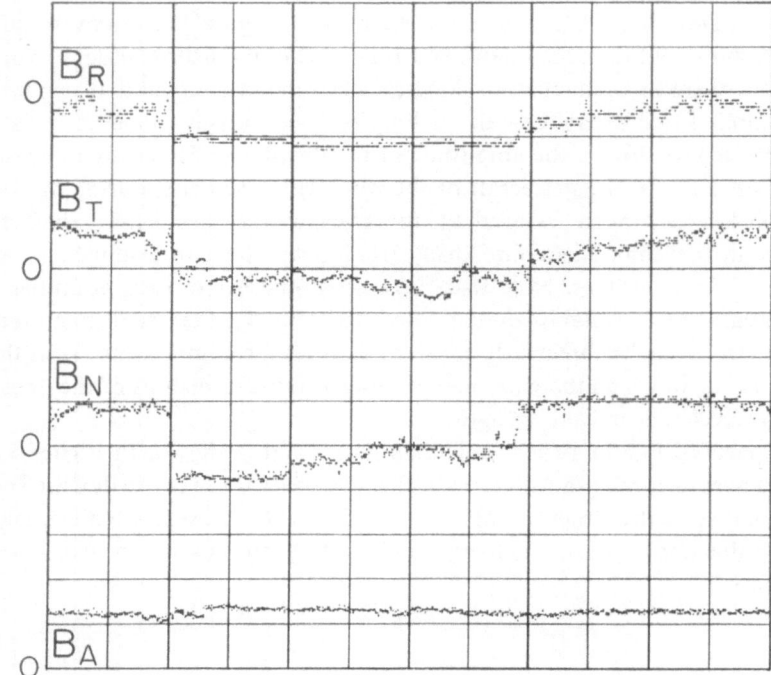

Fig. 4b.   Similar data during a somewhat disturbed time, hour 11 on the same day.

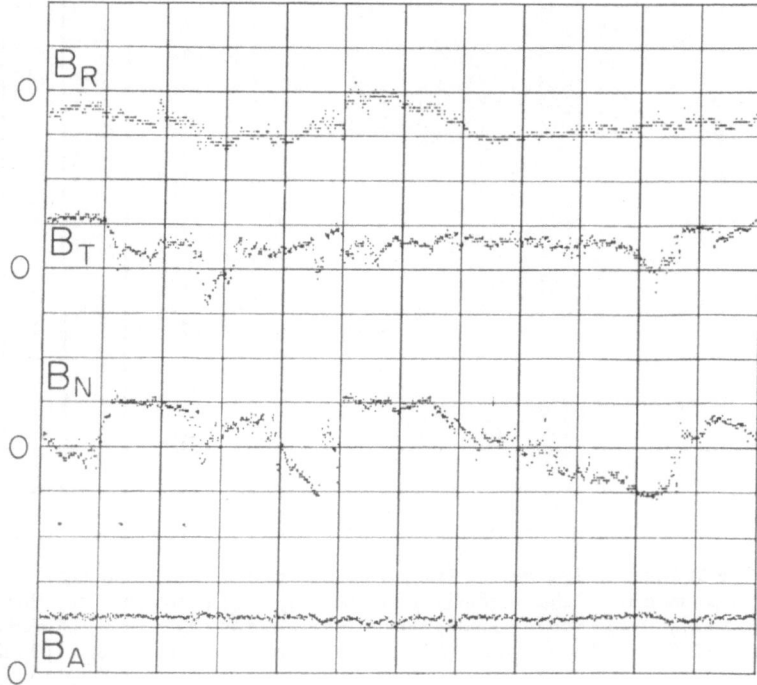

Fig. 4c.   Similar data during a quite disturbed time, hour 7 on the same day.

When data of this kind were first obtained eight or more years ago, it seemed very plausible that many of the features would propagate as Alfvén or magnetoacoustic waves. However, it proved very difficult to identify specific examples of such waves. Coleman (1967) found statistical evidence in the Mariner 2 data for outward propagating Alfvén waves. Unti and Neugebauer (1968) found one reasonably sinusoidal

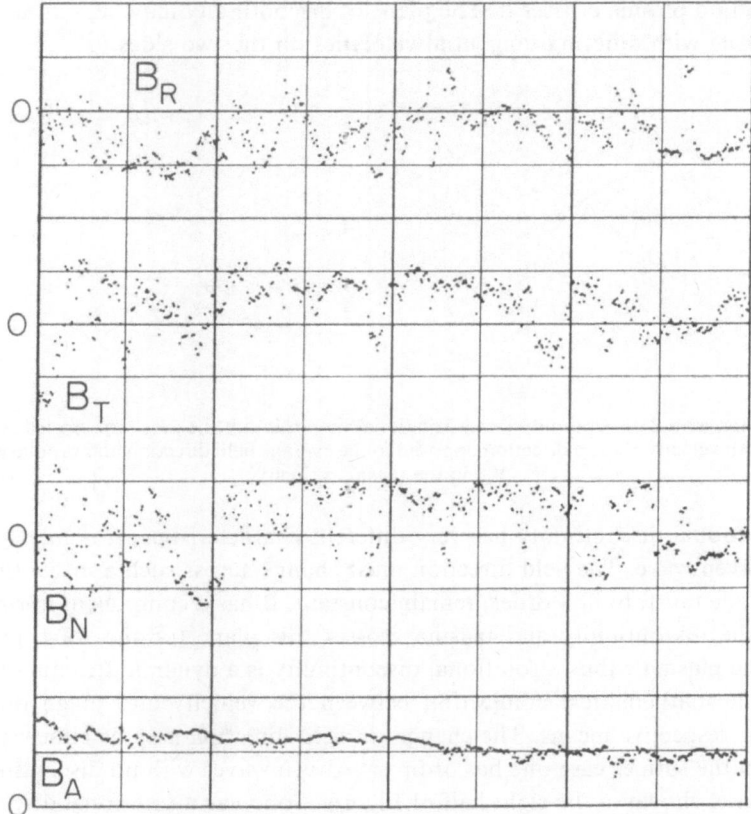

Fig. 5. Data similar to that in Figure 4 except that each point is an average over 2^{-9} day and the entire day 166 is shown.

example of such a wave. It was also expected that many of the discontinuities would be shocks, but it turns out that only a small fraction are. Shocks will be discussed in other papers at this meeting and hence will not be discussed further here.

If the discontinuities are not shocks, they should be either tangential or rotational discontinuities. Figure 6 may clarify the distinction. The left half expands a tangential discontinuity to the point where its internal structure is revealed. (For a true hydromagnetic discontinuity in a collisionless plasma, this means using dimensions small compared to the proton gyroradius; but many features that are appropriately treated

as discontinuities in the literature may have this appearance only because the scale
has been considerably compressed for convenience.) The magnetic field, **B**, changes in
magnitude or direction, or both, across a tangential discontinuity, but **B** is everywhere
parallel, or tangential, to the plane of the discontinuity. There is no necessity that the
field strength remain constant across the discontinuity, although it can. This discontinuity is a non-propagating structure that is convected with the bulk motion of the
plasma and no plasma crosses it. The plane of discontinuity may be, but need not be,
a shear plane with differing tangential velocities on the two sides.

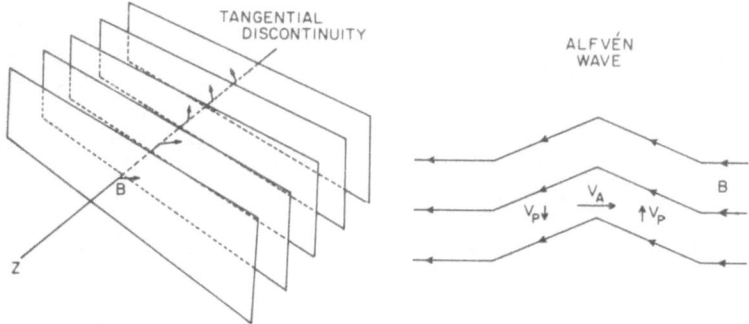

Fig. 6.   A tangential discontinuity and a rotational discontinuity; i.e., a sharp crested Alfven wave
traveling with velocity V_A in a direction opposite to the average field direction and producing changes,
V_p, in the plasma velocity.

The rotational discontinuity has very different characteristics; it is merely a sharp
crested Alfvén wave. The field direction must change across such a discontinuity but
its magnitude must, to first order, remain constant. **B** has a component normal to the
plane of the discontinuity and plasma crosses this plane (or the wave propagates
through the plasma); thus a rotational discontinuity is a dynamic structure and there
is a specific mathematical connection between the velocity and magnetic changes
about their respective means. The changes in field direction may be either smooth or
angular. In the former case one has ordinary Alfvén waves with no discontinuities, in
the later case, shown in the right half of Figure 6, one has a rotational discontinuity.

   Any discontinuity in the solar wind of either of these two types will make discontinuous changes in the components of the magnetic field of the kind so prominent near
the 2nd and 8th grid lines in Figure 4b; to distinguish between them plasma data are
needed. Since the plasma bulk velocity is so much larger than the Alfvén velocity in
interplanetary space, in either case the observed field changes are mainly due to the
convection of the solar wind past the spacecraft.

   Siscoe *et al.* (1968), Ness (1969), and Burlaga (1969) have shown that many of the
discontinuities are tangential and have shown that a number of the properties of the
interplanetary structure can be explained in terms of filamentary bundles of lines of
force that come all the way out from the Sun. For example, the data of Figure 4b
suggest that for 28 min the spacecraft was in one reasonably homogeneous bundle
(from the second nearly to the eighth grid line) and the rest of the time was in a homo-

geneous surrounding region. The fact that the field strength is the same in both requires that the plasma pressure is accidentally the same. The filaments are bounded by tangential discontinuities and are somewhat tangled. Sari and Ness (1969) have argued that most of the variability of the interplanetary magnetic field is due to such tangential discontinuities. Further discussion of these topics will be found in the review, later in this volume, by Burlaga.

Recently Belcher *et al.* (1969) have identified Alfvén waves using Mariner 5 data and have shown that non-sinusoidal, high amplitude, outwardly propagating Alfvén waves are a very common feature. They have urged that many of the discontinuities are sharp-crested Alfvén waves, i.e., rotational discontinuities. My own opinion is that, of the discontinuities, neither the rotational nor the tangential predominate strongly over the other type but that more of the variability of the interplanetary field is due to continuous changes than to discontinuities and that a substantial part of this variability is due to Alfvén waves.

Recent unpublished work by John Belcher at the California Institute of Technology provides even stronger evidence for this identification of these waves and explores further their relation to the high velocity streams. It is evident from the right half of Figure 6 that, as the wave propagates to the right, the perturbations in plasma velocity required if the plasma is to remain on the lines of force are correlated with the changes from the average of **B**. In the earlier work, only one component of this vector correlation was checked. Now, using Mariner 5 plasma data generously provided by Bridge *et al.* this has been extended to all three components. Figure 7 shows a good example.

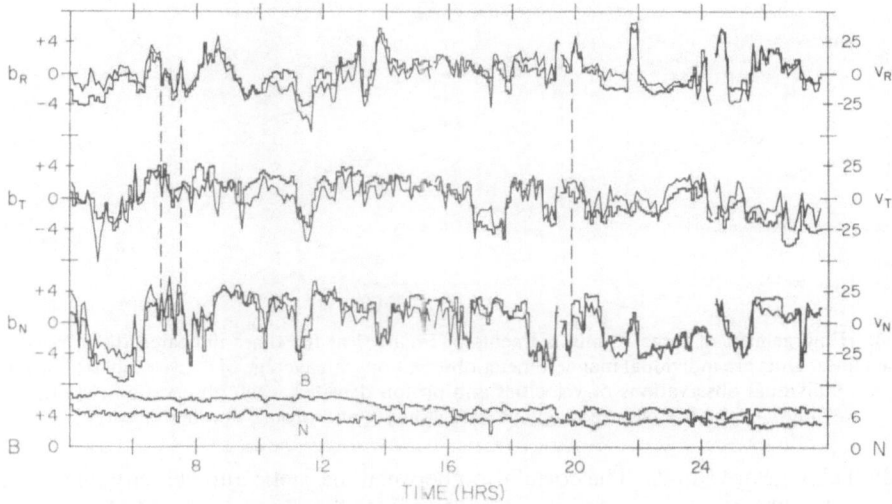

Fig. 7.   An example of good Alfvén waves. Proton number density N and the magnitude of **B** are shown by the two lowest curves. The remaining three pairs show the R, T, and N components of deviations from the average of the plasma velocity (curves with sloping segments, scale in km/sec on right) and of the magnetic field (curves with horizontal and vertical segments, scale in gammas on left). Each point plotted represents an average over 5.04 min. The data, taken from Mariner 5, extend over 24 hr on days 166 and 167 of 1967.

This particle density and magnetic field strength are reasonably constant, as expected for Alfvén waves. The correlations between magnetic and velocity fluctuations is very good indeed, again as expected. One sees, by studying Figure 6, that if the waves propagate in the same direction as the average of **B**, the magnetic and velocity fluctuations are anticorrelated while for propagation in the opposite direction, which is the case specifically illustrated, the correlation is positive. Since in Figure 7 the correlation is positive and since the field direction is inward toward the Sun, the waves must be propagating outward. The relative scales were so chosen as to make the correlation as good as possible, but the resulting ratio has just the value expected from the observed particle density if proper allowance is made for alpha particle contribution to the density and the effect of a modest pressure anisotropy on the standard theory of Alfvén waves. The agreement between theory and observation for this ratio, the good vector correlation, and the small variability of density and field strength make the identification as Alfvén waves seem very secure. It is also clear that during this interval there cannot be any substantial contribution from inward propagating Alfvén waves, magnetoacoustic waves, or static structures, since this would destroy the good correlation.

Figure 8 shows that some of these Alfvén waves are very sharp-edged and hence do correspond to rotational discontinuities. Two discontinuities and one relatively

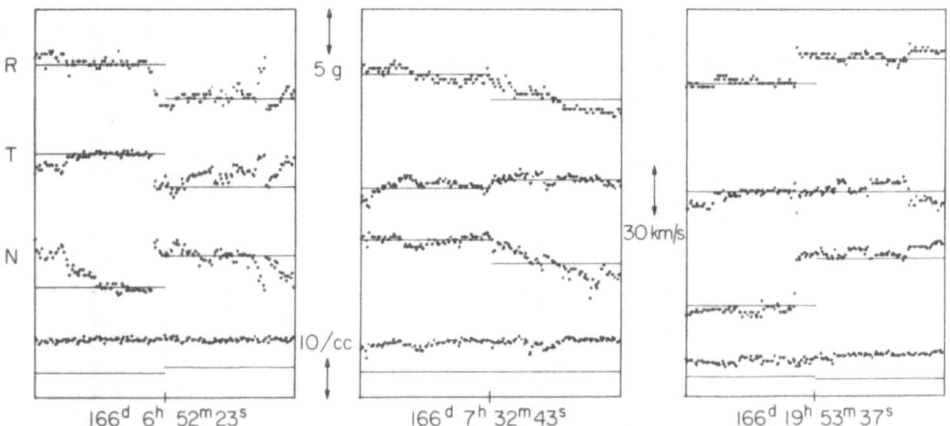

Fig. 8. Enlargement of three 10-min segments of Figure 7 at the times indicated there by vertical dashed lines. Dots are individual magnetometer observations, an average of 4.15 sec apart; horizontal lines are individual observations of velocities and proton densities, each observation requiring 5.04 min. At the bottom are magnetic field strength and proton number density.

gradual change are shown. The correlation between magnetic and velocity fluctuations is as good on this time scale as on the other. If the plasma properties could be measured as rapidly as the magnetic field, they would surely show as sharp a transition. It would take a remarkable coincidence, a real freak, for these discontinuities to be tangential but to have all the attributes required for a sharp-edged Alfvén wave and to be mixed in with ordinary Alfvén waves.

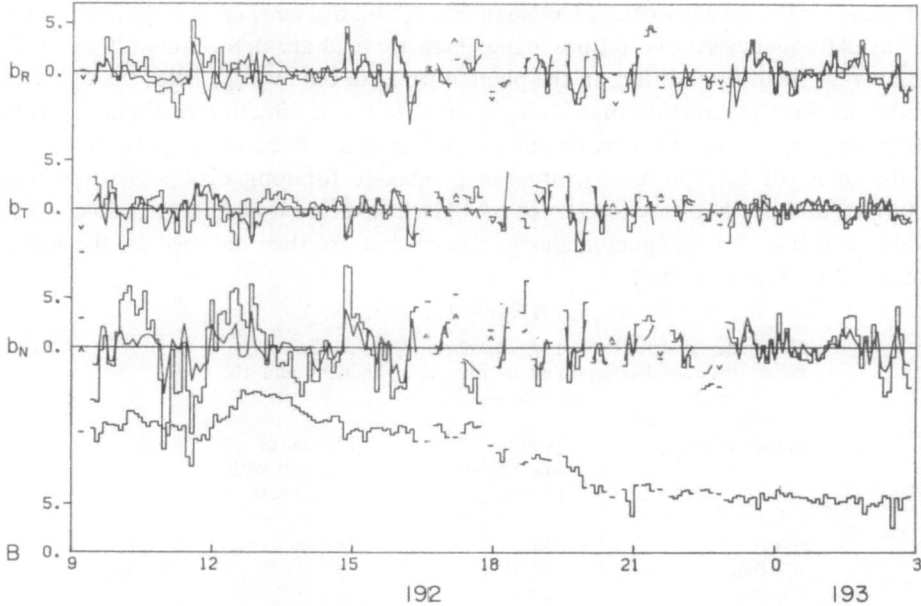

Fig. 9.   Mariner 5 data from day 192 and 193 showing waves of mixed character initially but becoming good Alfvén waves later. The curves are basically the same as in Figure 7 except that the curve
for proton number density is omitted.

Figure 9 shows that not all fluctuations are such pure Alfvén waves. Early on day
192 the field magnitude is changing and the correlations between magnetic and velocity fluctuations are poor. Presumably here there is a mixture of magnetoacoustic and
Alfvén wave modes Later B becomes more nearly constant and the correlations improve.

The data for the five-month Mariner 5 mission were analyzed to find what fraction
of the time, in 1967, outgoing Alfvén waves were clearly present. The correlation
coefficient between the radial components of the magnetic field and of the plasma
velocity were computed for each 6-hour interval for which adequate data were available. The first column of Table I lists ranges in which the absolute value of ϱ, the
correlation coefficient, might lie. The next column shows the number of intervals
found in each range. Nearly a third of the intervals show a correlation greater than
0.8 in absolute value and nearly two thirds show a correlation greater than 0.5. The
last column shows the fraction of the time that the correlation has the sign required
for outgoing waves. One sees that when the correlation is very high, the waves are
outgoing in essentially all cases; in fact, in each of the three apparent exceptions when
$|\varrho| > 0.8$, there are essentially no variations in the frequency range of the waves but
there are prominent linear trends in both variables. Cross spectral analysis shows that
when waves with coherences above 0.8 are present, they have significant amplitudes
for periods ranging at least from 10 min to 4 hr.

Extending the earlier work of Coleman (1967) and Siscoe *et al.* (1968), Belcher finds that at all frequencies the variations in the magnetic field are predominantly normal to the average field. In addition, in the plane normal to the average field, there is a 10% preference for the direction that is also normal to the R direction of Figure 1. Tentatively, we suggest that this may be due to the fact that Alfvén waves polarized normal to the plane of the spiral structure can propagate for long distances while wave polarized in this plane may tend to be partially converted by the large scale curvature of the field lines into magnetoacoustic modes that are then damped by the method suggested by Barnes (1966).

TABLE I

Frequency distribution of ϱ, the correlation coefficient between the radial component of the magnetic field and the plasma velocity

Range of $|\varrho|$	Number of intervals in this range	Percent of intervals with $\varrho\langle B_R\rangle < 0$
0.0–0.2	44	66
0.2–0.4	41	68
0.4–0.6	47	83
0.6–0.8	76	83
0.8–1.0	96	97
0.0–1.0	304	83

Many other interesting properties of these waves have been found, but there is space here for only a brief summary. Alfvén waves are most prominent in high velocity streams and their relation to the entire stream structure is of great interest. Figure 10 shows magnetometer and plasma data for a 7-day period during the first part of which the solar wind velocity, V_w, decreases as the spacecraft enters a low velocity stream. On day 236, V_w begins to rise as a new high velocity stream swings around with the Sun and overtakes Mariner 5. Typically, the density increases to a high value just before the velocity starts to rise and then drops abruptly early in the rise. Just at this time the temperature of the gas, measured by V_T, and the magnetic variability, σ_T increase abruptly. Plots similar to Figures 7 and 9 show that before this discontinuity Alfvén waves were not apparent; afterward they became very strong but not pure suggesting that either inward going waves or magnetoacoustic waves were also present. After the velocity increase levels off on day 237, the fluctuations are much more purely outgoing Alfvén waves, which continue through day 239. As shown by the lower value of σ_T, the amplitudes are not as large as in the region where the velocity was increasing rapidly. This whole pattern is typical, being found in many high velocity streams although at times Alfvén waves are found that are not associated with solar wind streams

There are two plausible models, either of which might explain the observations. The first assumes that the waves originate near the Sun, possibly largely along the inter faces between streams of different velocities or as remnants of the waves whose damping heats the corona. This model explains the predominantly outgoing character of the Alfvén very well but is, thus far, less successful in clarifying their detailed distribution across the high velocity stream. The second model assumes that the waves are generated continuously in the region where gas is compressed as a high velocity

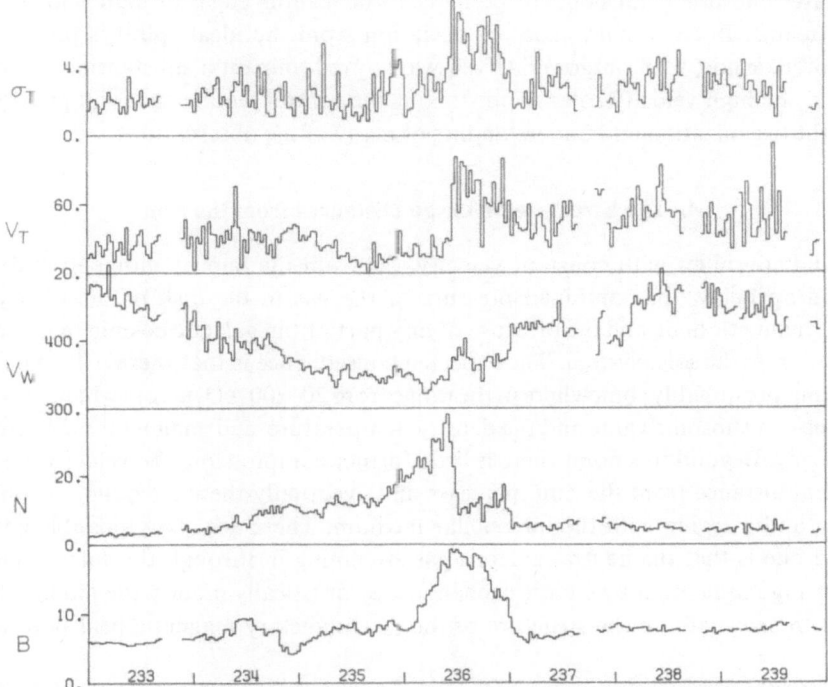

Fig. 10.   Averages of Mariner 5 magnetometer and plasma data for days 233 through 239. V_T is the most probable thermal velocity of a proton, assuming a Maxwellian distribution, σ_T is the square root of the sum of the variances of the three components of the magnetic field for 5.04 min intervals, and the other symbols have been defined in earlier figures.

stream overtakes normal low velocity solar wind as in the model discussed by Jokipii and Davis (1969). Both Alfvén and magnetoacoustic modes should be generated, but the latter would be damped by the Barnes mechanism. This explains the fact that the highest amplitudes and gas temperatures are found near the regions of highest density, but without further complication it does not explain the detailed distribution. Its main weakness is that unless the Alfvén waves are generated by means of an amplification process that selectively amplifies outgoing waves – and no plausible process of this kind has been suggested – there is no explanation for the preponderance of outgoing Alfvén waves. It may be that both mechanisms are involved since they are not mutually exclusive. To resolve these questions about the origin of the waves will probably require, and contribute to, a better understanding of the mechanisms that produce the velocity and density structure of the solar wind and its high velocity streams.

In summary, the features of intermediate scale in the structure of the interplanetary magnetic field include largely static filaments bounded by tangential discontinuities and convected outward by the solar wind. There are also some shocks associated with solar flares and other transients or perhaps at the leading edges of high velocity solar wind streams. But the main cause of deviation from the ideal spiral is propagating waves of all kinds, with outgoing Alfvén waves predominant a substantial part of the time and the high velocity solar wind streams the central feature that will provide the key to the organization and understanding of many of the observations.

4. The Structure at Large Distances from the Sun

If wind continues with constant velocity, far from the Sun its momentum density would drop below the combined pressure of the gas in the disk of the galaxy, the galactic magnetic field, and the pressure of any part of the galactic cosmic rays that are excluded from the solar system. The expected consequence is that there will be a shock transition, presumably somewhere in the range from 20–200 AU, across which the velocity drops to a subsonic value and the density, temperature and magnetic field strength rise sharply. Beyond this point there is little further compression, the velocity drops as the radial distance from the Sun increases and eventually there is cooling, if only by dilution and merging with the interstellar medium. There are many possible complications; one is that the neutral galactic gas streaming in through the solar wind and transferring momentum by charge exchange may drastically modify the model. It is of interest to speculate on the structure of the interplanetary magnetic field beyond the shock.

Between the Earth and the shock the radial component of **B** drops as r^{-2} while the tangential component drops only as r^{-1} so the spiral pattern becomes nearly circular. The fast streams and slow streams should all be merged by about 5 AU and the features associated with their interaction should fade away. It is possible that most of the waves will finally be damped out and most of the intermediate scale magnetic structure gradually disappear. However, the expansion in the T and N directions with no expansion in the R direction will continue to produce an anisotropic pressure tensor, and this might generate instabilities and magnetic irregularities on a scale of interest.

Passage through the shock should generate a great deal of magnetic noise such as is found in the Earth's magnetosheath. However, such noise might be damped out within a few tens of AU and the tubes of magnetic force return to a smooth spiral in which they are nearly hoops. If the plasma density remains constant, or increases as the hoops expand, the field strength will increase. Even inside the shock, the alternating magnetic polarity of adjacent solar wind streams will lead to a layered structure in which the field direction reverses from layer to layer, being mainly in the $\pm T$ direction. Inside the shock the layers will be of constant thickness, but outside the thickness will decrease with increasing radius. The increasing magnetic field strength means increasing currents in the interfaces between the layers. Since long times are involved, the effective electrical conductivity might be high enough that there would be some field line reconnection between adjacent layers.

This reconnection could produce a pattern like that is schematically indicated in Figure 11. Before reconnection, the lines are nearly normal to the direction of solar wind flow, spiraling slowly inward toward the Sun on the left side of the figure and outward to the Galaxy on the right. The uniform and stronger galactic field is shown at the top, outside the region of current interest. The reconnected field lines will make a series of U-shaped loops that eventually become closed. The tension in the lines of force adjacent to the loops produces azimuthal velocities that deflect V_w as indicated by the heavy arrows. Since on the average the right-hand side of a closed-loop island is likely to form first, being farther from the Sun, the azimuthal momentum imparted to the solar wind in this process is to the left, in the direction of the Sun's rotation.

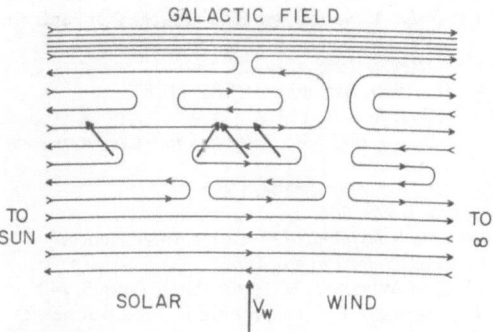

Fig. 11. Schematic representation of the magnetic field structure very far from the Sun if field line reconnection should occur.

This brings us back to a problem that was skipped over in the earlier discussion of the production of the azimuthal component of the solar wind velocity. The stresses in the magnetic field actually transport most of the angular momentum, more than does the azimuthal motion of the plasma. More of the slowing down of the rotation of the Sun is due to magnetic forces than to angular momentum imparted to the gas. In the usual mathematical treatment of the problem the boundary condition at infinity is, effectively, that the lines of force are anchored to something there. If we replace infinity by the interstellar medium, it becomes less clear what the solar lines of force can be anchored to. It could, of course, be the plasma in the single loop that was originally ejected from the Sun; but if Figure 11 represents the situation, it would be the plasma in all the smaller loops and islands. Angular momentum would be removed from the Sun and carried through the solar system mostly by magnetic stress and finally converted to the angular momentum of the azimuthal plasma motion at the interface with the galactic field.

Acknowledgements

I am greatly indebted to the Mariner 5 plasma experimenters, H. S. Bridge, A. J. Lazarus, and C. W. Snyder, and to the other Mariner 5 magnetometer experimenters, P. J. Coleman, D. E. Jones, and E. J. Smith for the use of carefully reduced data before publication. The work of J. W. Belcher in reducing, analyzing, and interpreting this data has made a major contribution to this review. Financial support from NASA under research grant NGR-05-002-160 is gratefully acknowledged.

References

Barnes, A.: 1966, *Phys. Fluids* **9**, 1483.
Belcher, J. W., Davis, L., Jr., and Smith, E. J.: 1969, *J. Geophys. Res.* **74**, 2302.
Bridge, H. S., Lazarus, A. J., and Snyder, C. W.: 1970, private communication.
Burlaga, L. F.: 1969. *Solar Phys.* **7**, 54.
Coleman, P. J., Jr.: 1967, *Planetary Space Sci.* **15**, 953.
Coleman, P. J., Jr., Davis, L., Jr., Smith, E. J., and Jones, D. E.: 1966, *J. Geophys. Res.* **71**, 2831.
Davis, L., Jr.: 1966, in *The Solar Wind* (ed. by R. J. Mackin, Jr. and M. Neugebauer), Pergamon Press, pp. 147 and 157.
Hundhausen, A. J.: *Space Sci. Rev.* **8**, 1968.
Jokipii, J. R. and Davis, L., Jr.: 1969, *Astrophys. J.* **156**, 1101.
Modisette, J. L.: 1967, *J. Geophys. Res.* **72**, 1521.
Ness, N. F.: 1969, *Proceedings of the 13th International Conference on Cosmic Rays*, Budapest, Hungary.
Ness, N. F. and Wilcox, J. M.: 1965, *Science* **148**, 1592.
Parker, E. N.: 1958, *Astrophys. J.* **128**, 664.
Parker, E. N.: 1963, *Interplanetary Dynamical Processes*, Interscience.
Sari, J. W. and Ness, N. F.: 1969, *Solar Phys.* **8**, 155.
Schatten, K. H., Ness, N. F., and Wilcox, J. M.: 1968, *Solar Phys.* **5**, 240.
Siscoe, G. L., Davis, L., Jr., Coleman, P. J., Jr., Smith, E. J., and Jones, D. E.: 1968, *J. Geophys. Res.* **73**, 61.
Snyder, C. W., Neugebauer, M., and Rau, V. R.: 1963, *J. Geophys. Res.* **68**, 6361.
Snyder, C. W. and Neugebauer, M.: 1966, in *The Solar Wind* (ed. by R. J. Mackin and M. Neugebauer), Pergamon Press, p. 25.
Unti, T. W. J. and Neugebauer, M.: 1968, *Phys. Fluids* **11**, 563.
Weber, E. and Davis, L., Jr: 1967, *Astrophys. J.* **148**, 217.
Weber, E. and Davis, L., Jr.: 1970, *J. Geophys. Res.* **75**, 2419.
Wilcox, J. M.: 1968, *Space Sci. Rev.* **8**, 258.
Wilcox, J. M. and Colburn, D. S.: 1970, *J. Geophys. Res.* **75**, 6366.
Wilcox, J. M. and Howard, R.: 1968, *Solar Phys.* **5**, 564.
Wilcox, J. M. and Ness, N. F.: 1965, *J. Geophys. Res.* **70**, 5793.

SCATTERING AND SCINTILLATIONS OF DISCRETE RADIO SOURCES AS A MEASURE OF THE INTERPLANETARY PLASMA IRREGULARITIES

V. V. VITKEVICH

U.S.S.R. Academy of Sciences, Moscow, U.S.S.R.

Abstract. It is well known that, as the radio-frequency radiation from distant discrete astronomica sources passes through the circumsolar (coronal) plasma and interplanetary plasma (extension of the corona, or the solar wind), it can undergo refraction and sometimes absorption by the (extended) corona as a whole and by larger-scale irregular structures (essentially regions of higher than average plasma density), scattering by smaller irregularities or cells, and scintillation. Observations at various wavelengths with radio-telescopes and interferometers (sometimes with networks of instruments) of the radio emission from the discrete sources after passing through the plasma (which the author calls the 'translucent' method) yield such quantities as the intensity I, characteristic scattering angle θ, scintillation index m, characteristic scintillation period τ_0, rms phase shift φ_0, and the transverse component of the velocity of motion of the diffraction pattern V. These quantities are functions of the time and location of the ray relative to the Sun (sometimes expressed as the angular distance or elongation ε of the source from the Sun, sometimes as the linear distance r); and all but the last are functions of the wavelength λ. With certain reasonable assumptions, parameters characterising the irregularities in the plasma structure can be derived, including a characteristic scale size a (or two orthogonal scale sizes a and b for those irregularities that apparently are elongated in shape), the drift velocity of the irregularities u, and the rms electron density fluctuation ΔN all as functions of heliocentric distances r or elongation ε, and the time, including the phase of the solar cycle or immediate level of solar activity. The results are not very sensitive to the assumptions that must be made concerning the thickness of the layer of plasma that dominates the production of the observed phenomena ΔL, the distance to the dominant region Z_0, the temperature T, etc.

This paper summarizes the results to date, and emphasizes in places that the method has not yet been fully exploited.

Typically, the characteristic scale size a derived from scintillation ranges from ≈ 30 km to close to the Sun to ≈ 300 km farther out, although the effects of selection imposed by the technique may be present. The rms electron density fluctuation ΔN, obtained from scattering and scintillation studies, are typically in the range 10^2–10^3 cm^{-3} close to the Sun, and smaller farther out, but corresponding to 1–10% throughout. The motions of the irregularities are predominantly radial, of the order of 400 km/sec in the ecliptic and apparently lower elsewhere. The irregularities apparently become radially elongated as they move away from the Sun. Assuming that the irregularities are preserved against dissipation by an interplanetary magnetic field, the field strength necessary to accomplish this is of the order of a few gammas or tens of gammas. Structures of larger size ($\approx 10^6$ km), greater density ($\approx 10^7$ cm^{-3}) and higher velocity ($\gtrsim 3 \times 10^3$ km/sec) are observed by other techniques.

1. Method

In this review are summarized the principal data on the plasma near the Sun and in the interplanetary medium that have been obtained by observations of radio emissions from discrete radio sources to which the plasma is 'translucent'. But first the method itself and principal steps in its development will be discussed.

The radio-astronomical method of investigation of the circumsolar plasma was developed in 1951 (Vitkevich, 1951), when only a few discrete radio sources were known. Among these was the Crab Nebula, a powerful source of radio emission. It

Dyer (ed.,, Solar Terrestrial Physics/1970: Part II, 49–66. All Rights Reserved.
Copyright © 1972 by D. Reidel Publishing Company, Dordrecht-Holland.

was noted that this source was near the ecliptic and that each year, in the middle of June, it passed close to the Sun. This led to the idea of exploring the solar corona by means of the 'translucent method', using radio waves coming through the corona from the Crab Nebula. In order to separate the radio emission of the Sun and the source, a radio interferometer having a lobe or fringe width of about 10′ to 30′ would be needed. In this case, we would have a good interference pattern for the source, because its angular size is about 5′, and a smoothed pattern for the Sun because its angular diameter is about 0.5° (Figure 1). In Figure 1, which is taken from the above-mentioned paper, the scheme of observations can be seen. As can be seen, it was

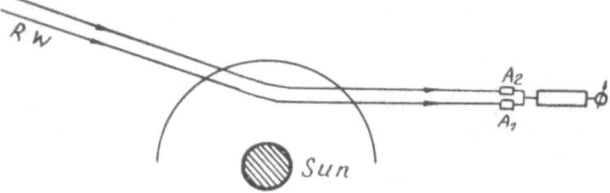

Fig. 1.   Schematic representation of the 'translucent' method of investigating circumsolar and inter-planetary plasma. The rays marked RW represent radio waves from a distant discrete radio source, here refracted by the circumsolar plasma and received at A_1, A_2, the linked antennas of a radio-inter-ferometer. The difference in direction of the incident ray and refracted ray is the scattering angle θ.

expected that refraction of radio waves in the solar corona would be observed and measured. Such observations at different wavelengths and with radio-interferometers with different resolving power were carried out at the Lebedev Physical Institute of the U.S.S.R. (Vitkevich, 1955) and at Cambridge, England (Hewish, 1955). These observations gave an unexpected result: the effects of great scattering of radiowaves in the outermost parts of the solar corona were found. These were attributed to small-scale irregularities in the circumsolar region. During the first few years, these irregularities were detected at distances up to 10 or 20 $R_\odot$ from the Sun and later, up to 50 or 60 $R_\odot$ by Australian radio-astronomers. For these investigations other discrete sources were used, as well as the Crab Nebula (Slee, 1961; Hewish, 1958; Högbom, 1960). The results of many years observations of the scattering of radio waves in the circumsolar plasma have been published: e.g., by Okoye and Hewish (1967), Vitkevich (1960a), Babii et al. (1965), Gorgolewski (1965), Hewish and Wyndham (1963), and Erickson (1964). But the radio sources used in these investigations had large angular sizes compared to those of the irregularities, and the outputs of the radiometers had large time-constants. Therefore we could get data only on the values of scattering angle; scintillations (variations in intensity) were not observed at that time.

When radio-telescopes with high sensitivity were developed and 'quasistellar objects' (QSO's or 'quasars' – radio sources having extremely small angular size) were discovered, it became possible to observe scintillations. This kind of observation was begun in 1964 in Cambridge (Hewish et al., 1964), and then spread to many other radio observatories.

The observation of scintillations has proved to be a much more sensitive method of investigating the irregularities of the interplanetary plasma, and it has become possible to investigate these irregularities up to elongations (angular distance from the Sun), equal to 180° (Antonova and Vitkevich 1966). Thus it has become possible to investigate the irregularities even at night, for plasma located more than an astronomical unit from the Sun.

In order to investigate the magnitude and direction of the velocity of the diffraction pattern in Cambridge (Hewish and Dennison, 1967; Dennison and Hewish, 1967) and in our laboratory (Vitkevich and Vlasov, 1966, 1968, 1969) three radio telescopes were used simultaneously.

It should be mentioned in passing that the discovery of pulsars has led to two new possibilities in the development of this method of cosmic plasma investigation. First because of the even smaller angular size of pulsars, it became possible to observe irregularities of even smaller angular dimensions. As a result, irregularities in the interstellar plasma that produce scintillations were found. Such investigations are now being widely developed.

Second, the periodicity of pulses coming from pulsars made it possible to measure the total number of electrons in the line of sight between the source and the observer. Such researches have been carried out for many pulsars; now we know much more about the average characteristics of the interstellar plasma. Undoubtedly, in the near future the solar corona will be also investigated by such methods (Vitkevich 1968).

Below (Sections 2–5) we shall discuss the problems of research on small-scale irregular structure, with a typical size of tens and hundreds of kilometres. In the conclusion (Section 6) we shall give some data on large-scale irregularities.

The basic characteristics of the interplanetary plasma structure can be described by the following parameters:

a (km)   – a characteristic physical scale-size of the irregularities, which are described by a function of the normal (gaussian) form,
$$R = \exp(-x^2/a^2).$$
ΔN (cm^{-3}) – the root-mean-square electron-density fluctuation;
a/b – the ratio of the average sizes of the irregularities or cells in the two directions perpendicular to the line of sight;
$\mathbf{u}$ (km/sec) – the drift velocity of the irregularities.

All these quantities vary with distance from the Sun; there are also a few data on their dependence on ecliptic latitude.

There are also some data about the dependence of the above-mentioned quantities on the phase of 11-yr cycle of solar activity, and on solar disturbances.

All parameters are determined using the theory of radio-wave propagation through an inhomogeneous plasma. This theory has been developed by many authors, and it is not necessary to give it here. Below are listed the quantities obtained or derived from observations. The principal relationships connecting them with parameters describing the irregularities will also be described.

θ (radian) – the 'scattering angle', or the characteristic angular width of the spreading or 'angular spectrum' of the received signal, corrected for the finite diameter of the source. It is measured with radio-interferometers having different resolutions. Such observations give fringes of different visibilities from which θ can be derived, assuming an angular spectrum of the form $I(\theta) = \text{const.} \exp(-x^2/\theta^2)$. If θ_0 is the angular size of the source itself, and θ_p the apparent angular size after scattering, then $\theta = \sqrt{\theta_p^2 - \theta_0^2}$.

For the observations of scattering the most important case is that in which $\varphi_0 \gg 1$, where φ_0 is the root-mean-square value of phase fluctuation φ in the line of sight.

m – the scintillation index, $m = \sqrt{(I-I_0)^2/(I_0^2)}$, in which I is the instantaneous intensity and I_0 is the mean intensity.

For investigations of the circumsolar plasma, the case, $\varphi_0 < 1$, is usually used. Then from the data on m, we can find φ_0, which involves a combination of ΔN and a (see below).

τ_0 – characteristic period of scintillation

$\mathbf{V}$ – the velocity and direction of motion of the diffraction pattern in the plane perpendicular to the line of sight.

To obtain $\mathbf{V}$, three radio-telescopes separated by reasonable distances are used simultaneously. Then the time of delay of arrival of the scintillations at successive sites gives the amount and direction of the velocity of diffraction pattern, from which the drift velocity of irregularities can be calculated. If for the same region of interplanetary plasma we also know the values of θ and φ_0, or of m, τ_0, and $\mathbf{V}$, we can derive all parameters of irregularities. The following relations are used.

For the scattering angle we have $\theta = \varphi_0 \lambda/\pi a$, ($\varphi_0 \gg 1$). Introduce the intermediaet expressions:

$$\varphi_0 = \pi^{1/4} k \Delta n \, (a \Delta L)^{1/2},$$
$$\Delta n = 4.5 \times 10^{-10} \, \lambda^2 \Delta N, \quad \text{and} \tag{1}$$
$$\varphi_0 = 3.75 \times 10^{-9} \, \lambda_m^2 \Delta N \, (a \Delta L)^{1/2},$$

in which Δn is the rms refractive index, k is the wave number, ΔL is the thickness of the layer in meters, $\Delta L/a$ is the number of irregularities along the line of sight, and λ_m is the wavelength in meters. By substitution and elimination, we obtain $\theta = 1.2 \times 10^{-9} \lambda_m^2 \Delta N (\Delta L/a)^{1/2}$, with λ_m, a, and ΔL all expressed in meters, and ΔN expressed in number per cm^3.

For the scintillation index m ($\varphi_0 \ll 1$), we have

$$m = \sqrt{2} \varphi_0 \left[1 - \frac{1}{1 + \dfrac{4\lambda^2 Z_0^2}{\pi^2 a^4}} \right]^{1/2}, \tag{2}$$

in which Z_0 is the distance to the layer. In the case of the distant regions, with

$Z_0 \gg a^2/\lambda$ we have $m = \sqrt{2}\varphi$. In the case of the near regions, with $Z_0 \ll a^2/\lambda$, we have

$$m = \sqrt{2}\varphi_0 \frac{2\lambda Z_0}{\pi a^2}. \tag{3}$$

For the investigation of the interplanetary plasma different wavelengths are used. Wavelengths of about 7 m or longer are used for studying the most distant regions from the Sun; decimetre and even centimetre wavelenths are used for studying the plasma regions nearest to the Sun.

2. Data on Scattering, Scintillations, and Root-Mean-Square Phase Deviations

Measurements of angular scattering have been carried out by many authors at different frequencies, with different baselines, in different years.

In Figure 2 the data for scattering angle plotted as a function of angular distance ε between the Sun and radio source (elongation of the source) are summarized. Maxi-

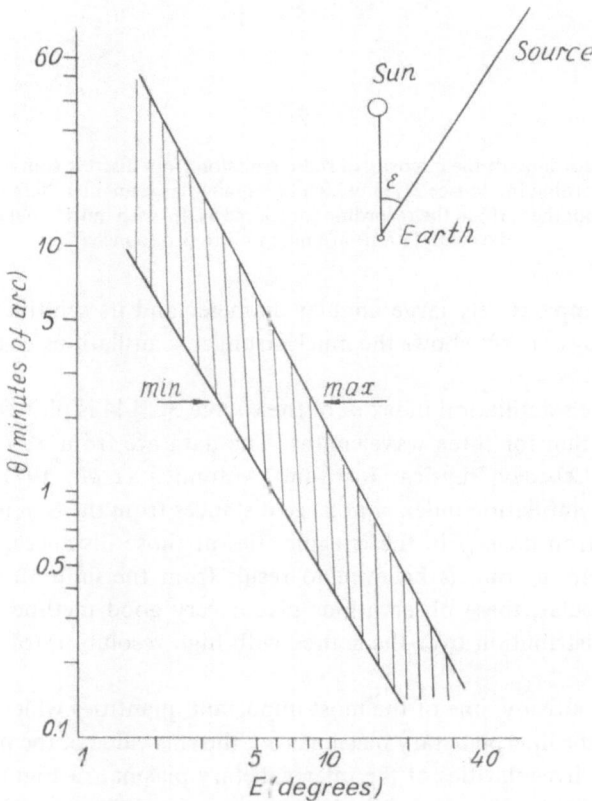

Fig. 2.   The range (minimum to maximum) of observed values of the scattering angle θ plotted against ε, the angle Sun-Earth-Source or 'elongation'. The values of θ have been normalized to wavelength $\lambda = 3.5$ m.

mum and minimum values of θ are shown; these have been normalized to a wave-length $\lambda = 3.5$ m, under the assumption that θ is proportional to λ^2. We can see that near the Sun the scattering angle θ is of the order of 10' arc-min to 1°. At $\varepsilon \approx 90°$, $\theta < 1'$. (Using the technique of very long-baseline interferometry, it will now be possible to extend these scattering observations up to $\varepsilon = 180°$, but no such data are yet available.)

Figure 3 is a copy of the observations of the intensity of two discrete sources, in which we can see the effects of scintillation. The source for the upper curve (source

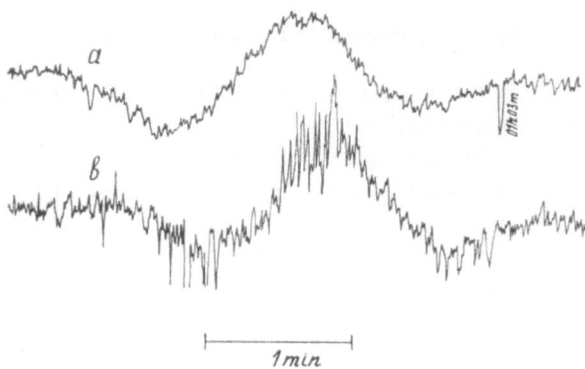

Fig. 3.   Typical recordings of the intensity of radio emission from discrete sources as functions of the time. (a) In the recording for source 3C33, which has a fairly large angular diameter, the scintillations are considerably smoothed; (b) in the recording for source 3C48, with much smaller angular diameter, the scintillations are much more pronounced.

3C 33) has a comparatively large angular diameter and its scintillations are greatly smoothed; the lower curve shows the much stronger scintillations of the small angular source 3C 48.

In Figure 4 the scintillation index m of the source 3C 144 is plotted against elongation ε from the Sun for three wavelengths. The data are from T. Antonova and A. Pynzar' at the Lebedev Physical Institute (Antonova et al., 1971). The observed decrease of the scintillation index m at large distances from the Sun is interpreted as a decrease of electron density in the irregularities at those distances. The decrease of scintillations near the Sun is believed to result from the finite angular size of the source. In particular, these observations give a very good method for investigating the brightness distribution over the source with high resolution (of the order of one second of arc).

As mentioned already, one of the most important quantities which characterize the irregularities of the interplanetary plasma is φ_0, the rms value of the phase fluctuations produced by the irregularities of the interplanetary plasma. In Figure 5 we have averaged the data obtained/from various investigators for different wavelengths as functions of elongation ε. Data for which $\varphi_0 > 1$ radian are usually used for the investigation of the scattering angle, and data for which $\varphi_0 < 1$ radian are used for scintilla-

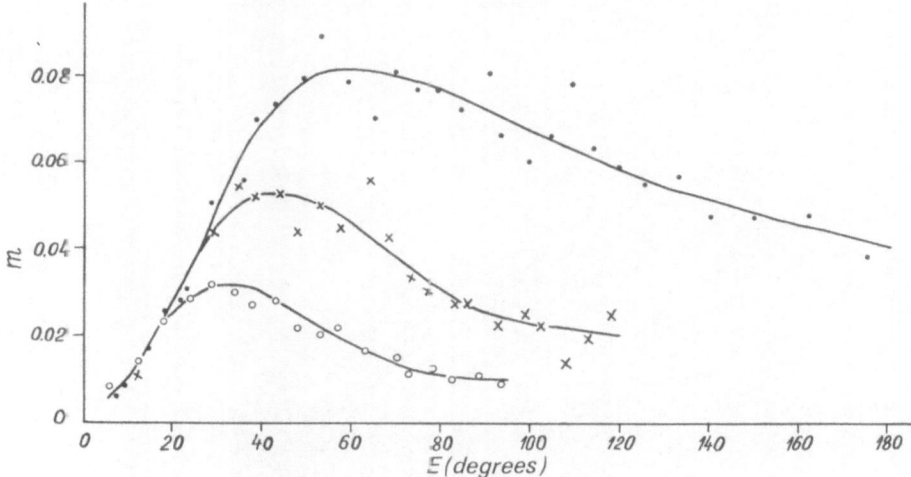

Fig. 4. The scintillation index *m* versus elongation ε, for three different wavelengths. Solid filled circles. $\lambda = 7.5$ m; Crosses, $\lambda = 5.0$ m; Open circles, $\lambda = 3.5$ m. (From observations of 3C144, Antonova *et al.* (1971).)

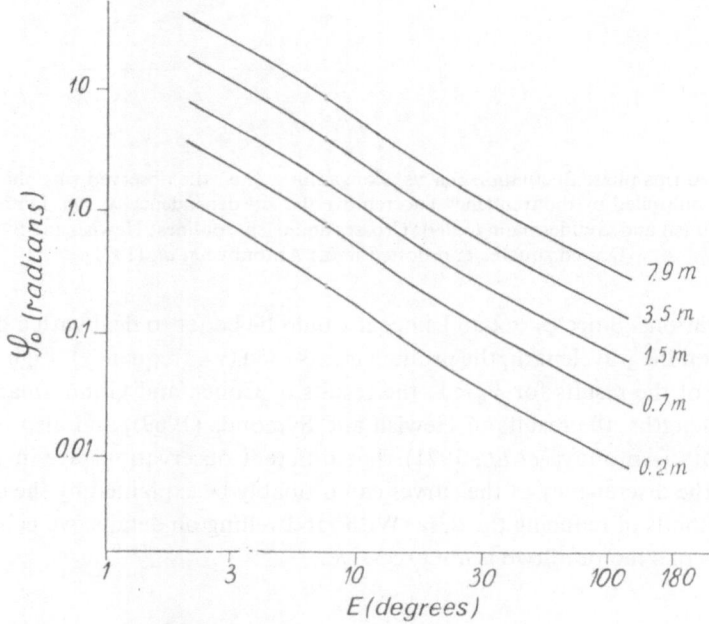

Fig. 5. Root-mean-square phase fluctuation φ_0 vs. elongation ε, for wavelengths in the range 0.2 m $\leqslant \lambda \leqslant 7.9$ m, as indicated at the right end of each curve.

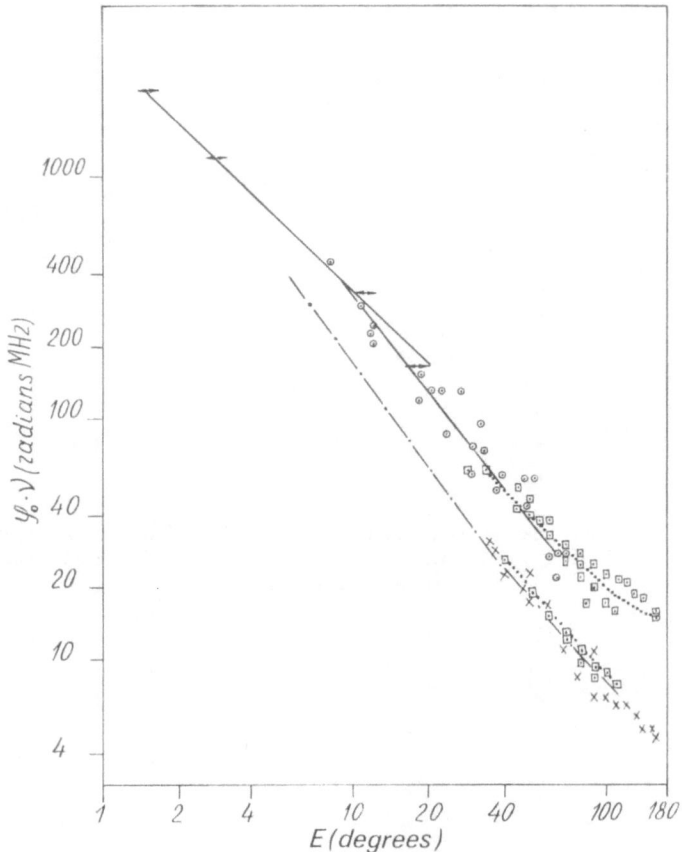

Fig. 6.   Scaled rms phase fluctuation $\varphi_0 \nu$ vs. elongation ε. (I.e., the observed rms phase fluctuation φ_0 has been multiplied by the frequency ν to remove the λ^{-1} dependence of φ_0). Dotted circles and solid lines: Cohen and Gundermann (1969); Crosses and dash-dot lines: Hewish and Symonds (1969; Dotted squares and dotted lines: Antonova *et al.* (1971).

tion observations. Since $\varphi_0 \propto \lambda$ and since it would be better to deal with a quantity not dependent on the wavelength, the product $\varphi_0 \nu$ is used (ν = frequency). Figure 6 summarizes many of the results for $\varphi_0 < 1$: the results of Cohen and Gundermann (1969) at short wavelengths, the results of Hewish and Symonds (1969), and also our Lebedev Institute data (Antonova *et al.*, 1971). The different observations are in good agreement, and the discrepancy of the curves can probably be explained by the difference in the two methods of reducing the data. Without dwelling on details, we believe that the discrepancy of a factor of two is not excessive.

3. Basic Parameters of Irregularities

The data on scattering, scintillations, and observations obtained at three sites give us the scale-size of the irregularities. These methods of investigation are more sensitive to small-scale irregularities. The scattering angle depends on the gradient of electron density in the irregularities, and large gradients correspond to small-scale irregularities. Thus our information on small-scale irregularities is better than for the larger-scale ones.

The scintillation index m depends strongly on the relation between the distance to the irregularities Z_0, their size a, and the wavelength λ. In order to have sufficiently strong scintillations, the inequality $a^2 < \lambda Z_0$ must be satisfied. With $a^2 > \lambda Z_0$, the scintillation index m becomes very small. From these inequalities we can determine the maximum values of a that can produce measurable scintillations.

In Figure 7, the values of a are plotted as a function of the elongation ε. The upper limit marked by arrows is that calculated from the inequality above and gives the

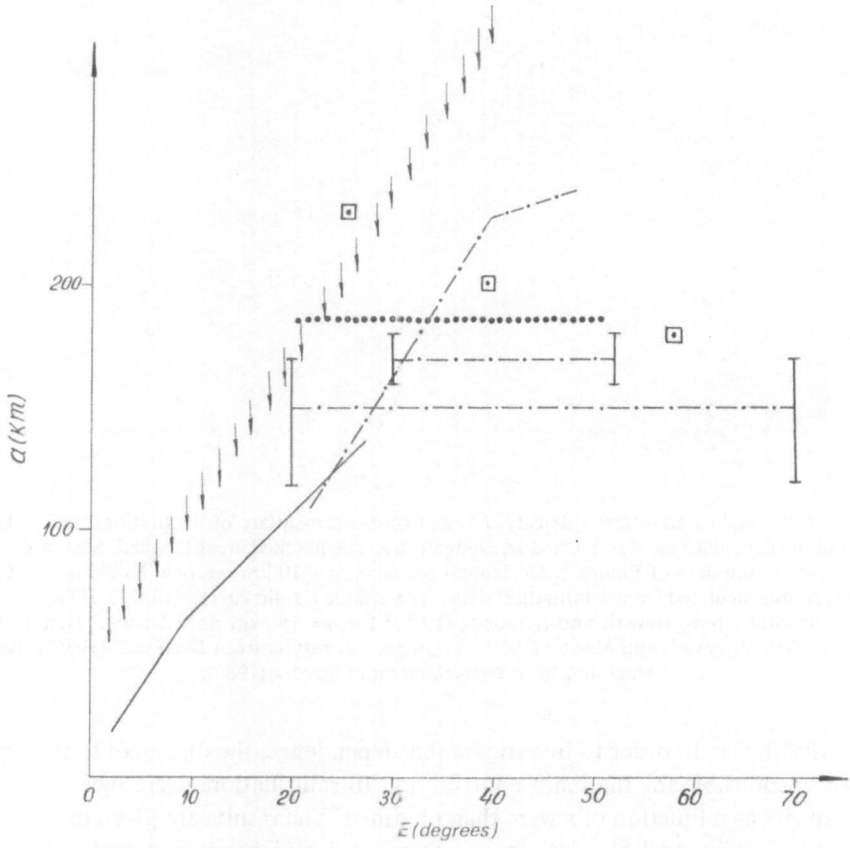

Fig. 7. Scale size a of plasma irregularities *vs.* elongation ε. The arrows indicated the upper limit of a that can be determined by the scintillation method, and thus show the limitations of the method. Solid curve: Cohen and Gundermann (1969); Dash-dot lines: Hewish and Symonds (1969), Dennison and Hewish (1967), and Hewish *et al.* (1966); Dotted line and squares: Vitkevich and Vlasov (1969).

maximum possible values of a that can be reached by this method of investigation. In this figure we can see that the quantity a derived from observations tends to decrease toward the Sun; it is not excluded that this decrease is in part the result of selection imposed by the limitations of the observational method, as just described.

From data on scattering or scintillation and on the distribution of irregularity scale-size a, we can get the rms electron density of the irregularities. Most of the available data are summarized in Figure 8a, plotted as a function of ε. We are also interested in the dependence of the rms electron density fluctuation ΔN on the helio-

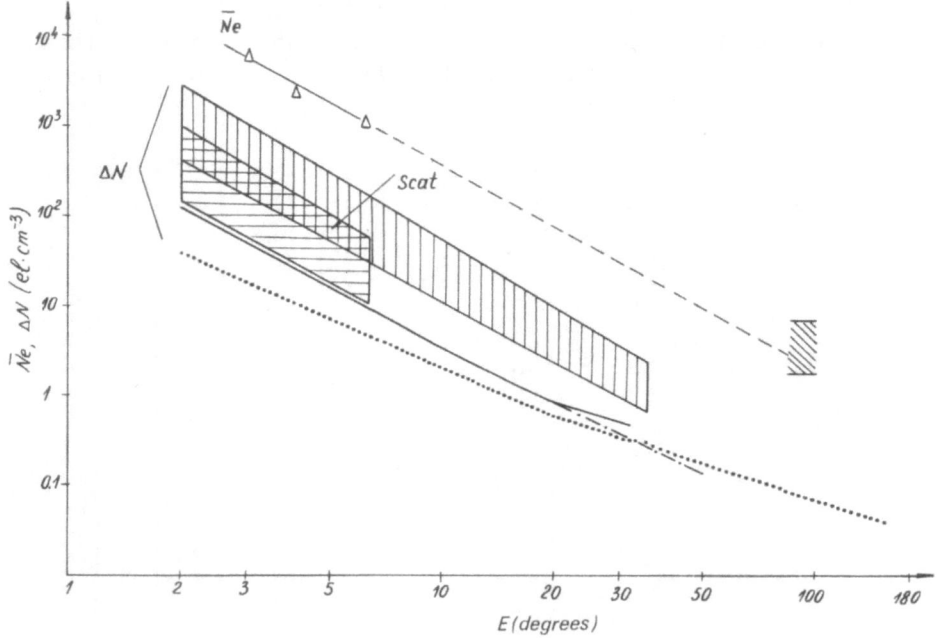

Fig. 8a.   Observed mean electron density, $\bar{N}_e$, and root-mean-square of fluctuations in the electron density in the irregularities, ΔN, plotted vs. elongation ε. The hatched areas labelled 'Scat' are derived from the scattering data of Figure 2. Horizontal hatching, $a = 10$ km; vertical hatching, $a = 90$ km. The curves are calculated from scintillation data: for variable a solid curve, Cohen and Gundermann (1969); dash-dot curve, Hewish and Symonds (1969); for $a = 180$ km dotted curve, Dennison and Hewish (1967), Vitkevich and Vlasov (1969). Triangles, N_e results from Blackwell (1956); diagonal hatching, from Neugebauer and Snyder (1966).

centric distance r. In order to investigate this dependence, the distances to the regions that are responsible for the main contribution to scintillations were derived and the values of ΔN as a function of r were then obtained. The results are given in Figure 8b.

As we see in Figure 8, the data from scattering give higher values of ΔN than those from scintillation. This is probably explained by the difference in scale-size for the irregularities which are chiefly responsible for scattering and scintillations, respectively.

It is necessary now to note that an anisotropy or ellipticity in the shapes of irregu-larities has been observed. Radio-interferometer measurements have shown that the scattering angle, and hence effective diameter of the irregularities, are functions of position angle (or direction in the plane perpendicular to the line of sight). If the scattering angle is plotted as the radius (in polar coordinates) against position angle,

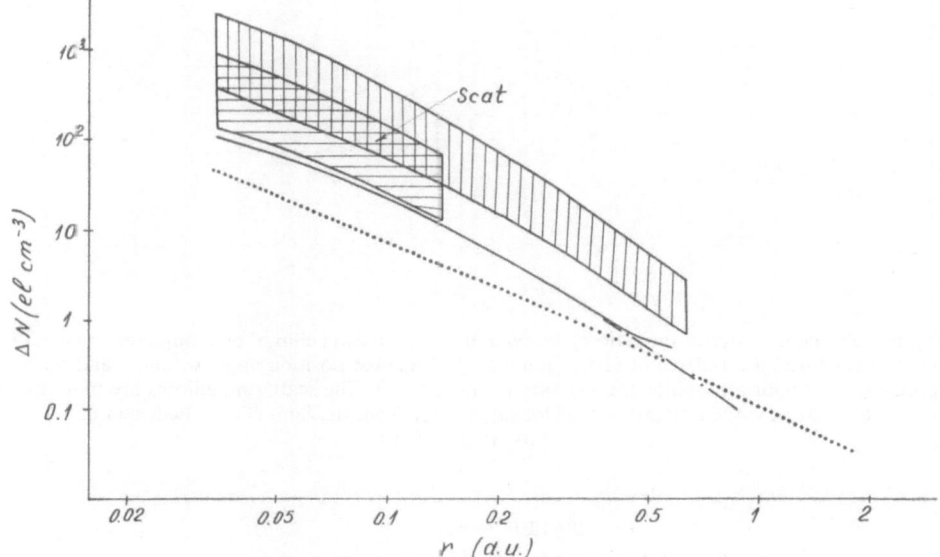

Fig. 8b.   Rms electron density fluctuation ΔN vs. heliocentric distance r. (Symbols as in (a).)

an elliptical figure results. Usually the minor axis of the scattering ellipse is oriented approximately along the projection of line from the Sun to the irregularity. Since the scattering angle θ is inversely proportional to a, this means that the irregularities are elongated mainly in the radial direction. From this an important conclusion follows, that there must be some restraint acting against dissipation of the irregularities that has a predominant radial component. The observational data shown in Figure 9 show that b/a near the Sun is about 0.8 but decreases at $\varepsilon > 5°$ to a more or less con-stant value of about 0.5, i.e., the scattering ellipse is more nearly circular near the Sun, becoming more elongated at greater distances. From this we conclude that the irregularities are more nearly spherical closest to the Sun and become more elongated ellipsoids farther out.

The presence of small-scale irregularities, as well as their elongated shape, raises a question about the nature of the agent that restrains them from dissipation. The con-ventional point of view is that magnetic fields (a regular component as well as irre-gular) are responsible. If that is the case, the magnetic fields must be strong enough that the gyromagnetic radius of the particle (proton) motion will be less than, or of the order of, the irregularity scale-size. From the expression for the radius of gyration

$r_H = mcv/eH$ we can find a lower limit H_{min} of the magnetic field strength for different
values of the proton temperature T, from which the particle velocity v relative the
bulk motion can be derived. Table I shows the minimum field strengths calculated for
several values of T and r_H.

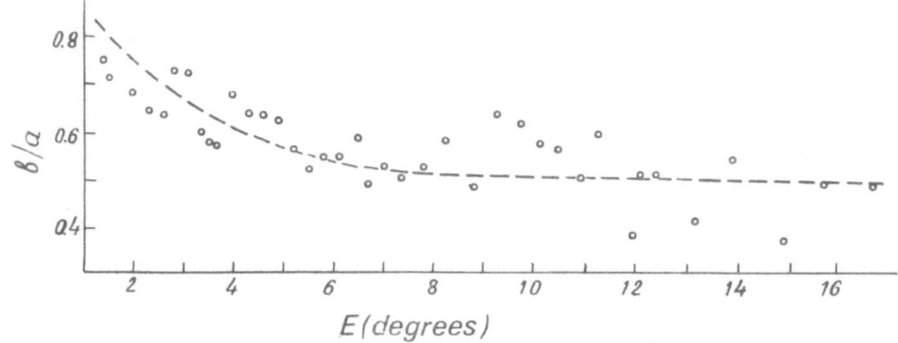

Fig. 9.   The dependence of the ratio of axes b/a of the 'scattering ellipse' on elongation angle ε. (The
'scattering ellipse' is a polar plot of scattering angle θ against position angle ψ, and a and b are respectively the major and minor (semi-) axes of this ellipse). The scattering ellipses are more nearly
circular near the Sun and more elongated away from the Sun. (From Bedevkin and
Vitkevich (1967).)

TABLE I

Magnetic fields (in gammas) for different gyroradii r_H
and plasma temperatures T.

T r_H	30 km	100 km	300 km
10^6	20	7	2
2×10^5	10	3	1

4. Velocities and Directions of the Motions of Irregularities

Now let us discuss some results of the measurements of the drift of irregularities
obtained by using a net of three radio-telescopes. If we have simultaneous scintillation
observations from three spaced radio-telescopes and can determine the time delay in
the arrival of fluctuations between stations, we can find the direction and magnitude of
velocity of the diffraction pattern.

   The direction of motion on the average is radial with respect to the Sun. Data obtained during 1967–1969 at the Lebedev Physical Institute by V. Vlasov (private
communication) are given as a histogram in Figure 10. We see some deviations from
the radial direction which evidently cannot be explained by errors of observation.
In Figure 11, taken from the paper by Vitkevich and Vlasov (1969), the results seem to
show a flow of plasma deviating away from the radial direction towards the ecliptic

plane within the range of elongation $25° \leqslant \varepsilon \leqslant 50°$. This important question requires further investigation.

Data on the velocity of the diffraction pattern from observations of the source 3C 48 are given in Figure 12. For this ecliptic latitude ($\approx 20°$) the velocities on the average are close to 300 km/sec. Incidentally, no noticeable dependence of V on ε was observed. The observations obtained by Dennison and Hewish (1967) using the source 3C 144, which is near the ecliptic plane, yielded magnitudes of the drift velocity that are 40% larger (see Figure 13). Velocity data using three observation sites are presented in Table II.

One should remember that, for radio-astronomical observations, it is the velocity of

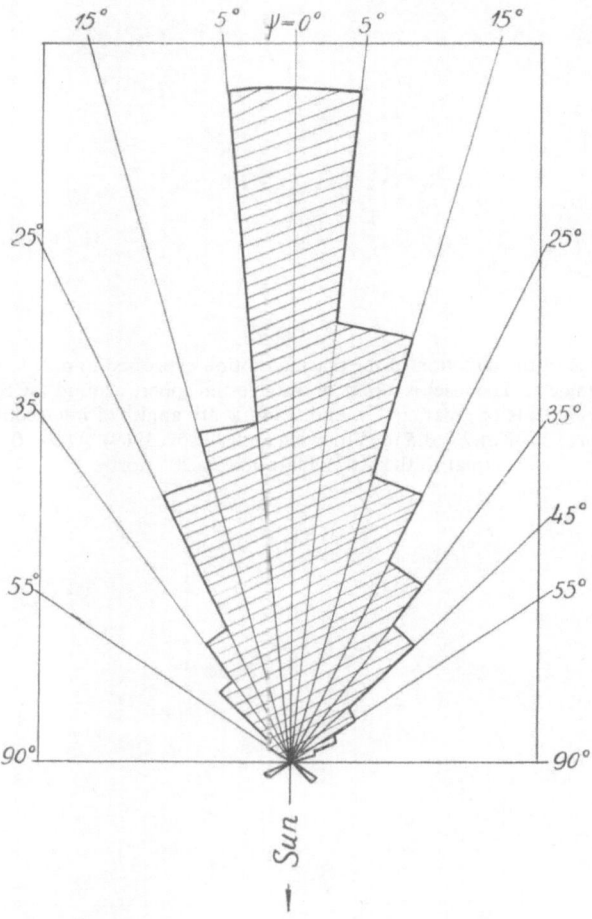

Fig. 10.   Histogram (in polar coordinates) of the preferential directions of the motion of plasma irregularities. The independent variable ψ is the position angle measured from $\psi = 0$ which corresponds to the direction away from the Sun along the great circle passing through the Sun and source. (See inset in Figure 11 for the geometry.)

the diffraction pattern motion that is directly measured. Two factors should be taken into account in order to reduce this observed velocity to the velocity of plasma motion **u**: (1) Measurements cover a certain region along the line of sight, with a thickness ΔL. With the model of radial outflow, the measured velocity **V** is smaller than the true velocity of irregularities **u**, by an amount due to projection which varies within the range of thickness ΔL. (2) The direction of the normal to the front of the diffraction pattern may not coincide with the direction of the velocity. If the angle between the two directions is ξ, we would obtain values of $V \cos\xi$.

In respect of both these factors, the true velocities of the irregularity motions

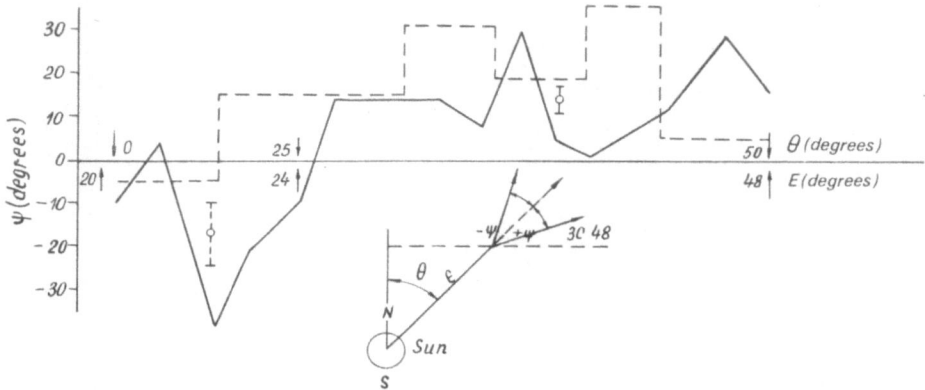

Fig. 11.   Mean value of the direction of the plasma motion expressed in position angle ψ as a function of elongation angle ε. (The inset is a part of the celestial sphere around the Sun and source, segments ε and N are segments of great circles, and θ and ψ are angles of intersection of great circles.) The data are for source 3C48 at $\lambda = 3.5$ m (Vitkevich and Vlasov, 1969). At $\theta = 0$, ε is a minimum and equal to the ecliptic latitude, $\approx 20°$ north.

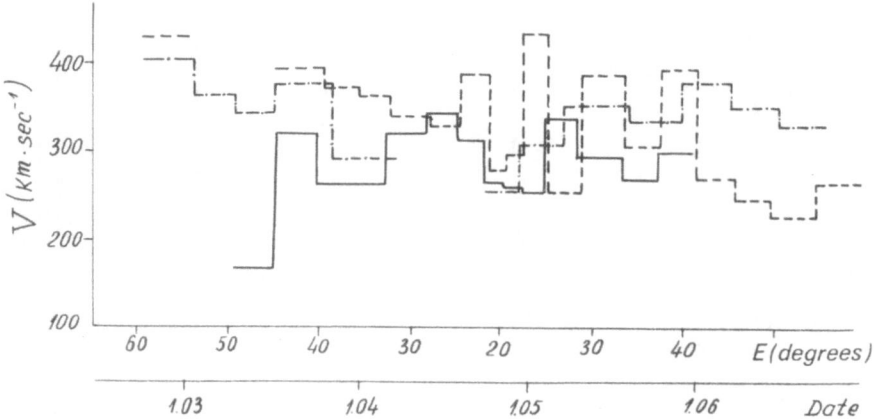

Fig. 12.   Dependence of the velocity V of the diffraction pattern on elongation ε for the source 3C48 at ecliptic latitude $\approx 20°$ north. Dash-dot lines: Dennison and Hewish (1967); Solid lines: Vitkevich and Vlasov (1969); Broken lines: unpublished 1968 data of the Lebedev group.

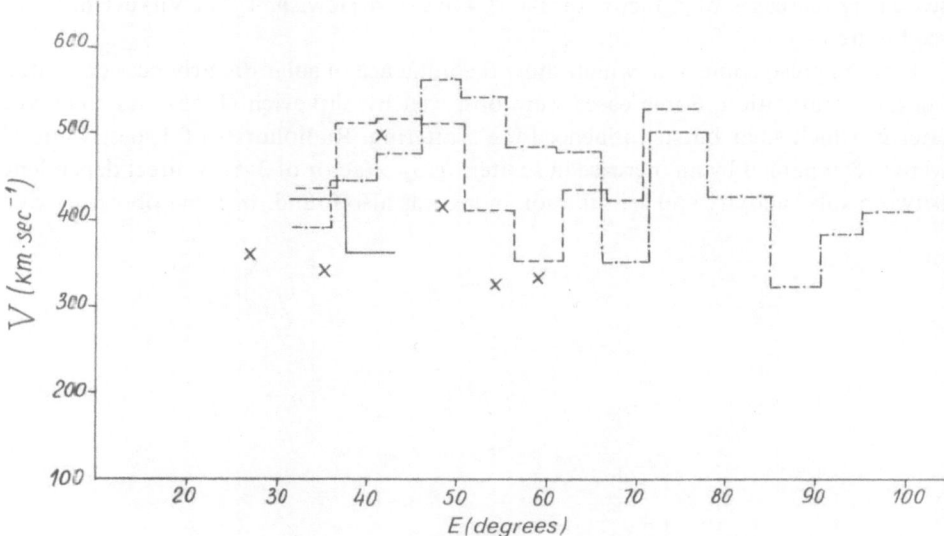

Fig. 13. Dependence of the velocity V of the diffraction pattern on elongation for the source 3C144, close to the ecliptic plane. The velocities appear to be systematically higher than in Figure 12. Crosses: Hewish and Symonds (1969); Solid lines: Vitkevich and Vlasov (1969); Broken lines: unpublished 1968 data of the Lebedev group; Dash-dot lines: unpublished Lebedev data, 1969.

TABLE II

The mean values of the diffraction pattern velocities (km/sec)

Source	3C48		3C144	
Year	Dennison and Hewish (1967)	Vitkevich and Vlasov (1969)	Dennison and Hewish (1967)	Vitkevich and Vlasov (1969)
1967	275 ± 70	260 ± 55	370 ± 90	390
1968	—	325 ± 150	—	440 ± 100
1969	—	280 ± 95	—	430 ± 90
Heliocentric latitude	$20°$–$60°$ North		$\sim 0°$ (Ecliptic)	

would be somewhat higher than the observed velocities of the diffraction pattern motion. The corresponding total correction factor has not been calculated thoroughly in all cases, but it may be as much as 1.3.

5. The Effect of Solar Activity on Scattering and Scintillation

Let us now examine the dependence of the parameters of small-scale irregularities on the phase of the 11-yr solar activity cycle. Observations of scattering were made systematically over a full cycle. It was found that during maximum solar activity, scattering increases by a factor of 3–6 (Okoye and Hewish, 1967; Vitkevich, 1960a; see Figure 14.)

There are also some data which show the influence of solar disturbances on scattering and scintillation. Some cases were observed by Vitkevich (1955) and some year later in which solar bursts influenced the scattering. Radiobursts of Types II and III were accompanied by an increase in scattering by a factor of 3–5. A direct dependence between solar activity and scintillation index was also found. In 1966, observations at

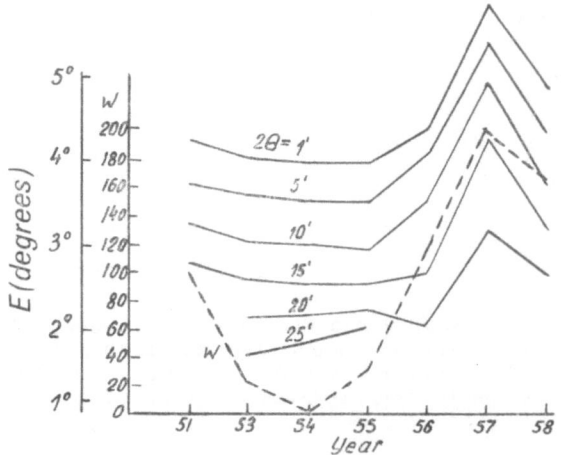

Fig. 14.   Amplitude of scattering as a function of phase of the solar activity cycle. Each solid-line curve is labelled with a corresponding value of the (double) scattering angle 2θ; when read against the ε scale (outer scale at left) it gives the elongation corresponding to that value of the scattering angle for the year in question (yearly averages). The Wolf sunspot number index is given (broken curve labelled W and inner scale at left) as a measure of the level of solar activity. It can be seen that the whole distribution pattern of scattering (larger amplitude closer to the Sun, decreasing amplitude with greater elongation) expands and contracts in phase with solar activity.
(Based on Vitkevich (1960a).)

Arecibo Observatory showed that the scintillation index increased 4–8 times during Type II flares on the solar disk (Sharp and Harris, 1967). Furthermore, the Cambridge workers have found that the scintillation index varies with a periodicity of about 27 days at $\varepsilon \leqslant 40°$ and the ecliptic latitude $\leqslant 30°$ (Dennison and Hewish, 1967), which appears to be the effect of active regions being brought into position by solar rotation. More systematic data about the effect of solar activity on small-scale irregularities are needed to investigate more throroughly this most important problem.

6. Large-Scale Irregularities

The 'translucent' method gives not only detailed information about small-scale structures, but also some information about large-scale formations. On this subject, we have only very limited information from a few single observations.

Two types of the cases are observed. The first type is somewhat analogous to the scintillations already described. It is manifested as a disturbance of the fringe pattern in the recording of radio emission from the source. Vitkevich (1955) gives examples of disturbances lasting typically 1–2 min at distances of about 15 $R_\odot$ from the Sun. This phenomenon was confirmed later by Slee (1959). Indications are that a rapidly moving plasma, flowing with a velocity of the order of 10^3 km/sec, can cause such disturbances. Vitkevich (1960b) examined the possible causes of the effects described above. He determined that the velocity of motion of the plasma formations responsible for the effects must be 3×10^3 km/sec and more. The estimate of the plasma parameters, e.g. electron density, depends on the mechanism that is assumed to produce the decrease in radio intensity. The typical size a of such large irregularities appears to be 0.5×10^6 to 1×10^6 km. Using this value and assuming that the effect is due to absorption, we can calculate the electron density for different temperatures T, as shown in Table III.

TABLE III

Electron density of large-scale irregularities

$T(^\circ K)$	10^6	10^5
$N(\text{cm}^{-2})$	$(1.5 \text{ to } 5) \times 10^7$	$(2.5 \text{ to } 8) \times 10^6$

If the decrease in intensity is due to the 'lens' effect, we can again estimate the parameters. When such a lens covers the source as it moves, the intensity is reduced. It appears that, with $a = 0.5 \times 10^6$ km, electron densities 1.5 to 7×10^6 cm^{-3} are required. At the present time, it is not possible to decide which case we are dealing with; systematic investigations of these seldom observed phenomena are necessary.

Another effect through which the presence of the large-scale irregularities is displayed is the noticeable but seldom observed refraction of radio waves. Such cases have been observed at $\lambda = 5.8$ m and $\lambda = 3.5$ m, using three radio-interferometers simultaneously. Vitkevich (1958) obtained the following results. At a distance up to 10–15 $R_\odot$ and at a wavelength of 5.8 m, the refraction was about 1°; at a wavelength of 3.5 m during the observations in 1963 refraction was 1.5–4 min of arc at distances of 30–40 $R_\odot$ from the Sun. These results are explained by the presence of individual large-scale inhomogeneities. For the first case the excess of electron density was $\Delta N/N_0 \approx 10$; for the second case $\Delta N/N_0 \approx 10^2$. These results point to the possibility of advancing the study of large-scale structures by this method.

Acknowledgements

The author takes this opportunity to thank the staff of the Radio-Astronomy Laboratory of the Lebedev Physical Institute for their help in the preparation of this review, especially T. D. Antonova and A. Pynzar'.

References

Antonova, T. D. and Vitkevich, V. V.: 1966, *Astron. Tsirkul. U.S.S.R.*, **385**.
Antonova, T. D., Panadjan, V. G., and Pynzar', A. V.: 1971, *A. J. U.S.S.R.* **48**, 1.
Babii, V. I., Vitkevich, V. V., Vlasov, V. I., Gorelova, M. V., and Sukhovei, A. G.: 1965, *A. J. U.S.S.R.* **42**, 107.
Bedevkin, V. F. and Vitkevich, V. V.: 1967, *Trudy PhIAN U.S.S.R.* **38**, 96.
Blackwell, D.: 1956, *Monthly Notices. Roy. Astron. Soc.* **116**, 56.
Cohen, M. H. and Gundermann, E. J.: 1969, *Astrophys. J.* **155**, 645.
Dennison, P. A. and Hewish, A.: 1967, *Nature* **213**, 343.
Erickson, W. C.: 1964, *Astrophys. J.* **139**, 1290.
Gorgolewski, S.: 1965, *Acta Astron.* **15**, 261.
Hewish, A.: 1955, *Proc. Roy. Astron. Soc.* **A228**, 238.
Hewish, A.: 1958, *Monthly Notices Roy. Astron. Soc.* **118**, 534.
Hewish, A. and Wyndham, J. O.: 1963, *Monthly. Notices Roy. Astron. Soc.* **126**, 469.
Hewish, A., Scott, P. F., and Wills, D.: 1964, *Nature* **203**, 1214.
Hewish, A., Dennison, P. A., and Pilkington, J. D. H.: 1966, *Nature* **209**, 1188.
Hewish, A. and Dennison, P. A.: 1967, *J. Geophys. Res.* **72**, 1977.
Hewish, A. and Symonds, M. D.: 1969, *Planetary Space Sci.* **17**, 313.
Högbom, J. A.: 1960, *Monthly Notices Roy. Astron. Soc.* **120**, 530.
Neugebauer, M. and Snyder, C. W.: 1966, *J. Geophys. Res.* **71**, 4469.
Okoye, S. E. and Hewish, A.: 1967, *Monthly Notices Roy. Astron. Soc.* **137**, 287.
Sharp, L. E. and Harris, D. E.: 1967, *Nature* **213**, 377.
Slee, O. B.: 1959, *Australian J. Phys.* **12**, 234.
Slee, O. B.: 1961, *Monthly Notices Roy. Astron. Soc.* **123**, 223.
Vitkevich, V. V.: 1951, *Dokl. Akad. Nauk U.S.S.R.* **77**, 585.
Vitkevich, V. V.: 1955, *Dokl. Akad. Nauk U.S.S.R.* **101**, 429.
Vitkevich, V. V.: 1958, *A. J. U.S.S.R.* **35**, 52.
Vitkevich, V. V.: 1960a, *A. J. U.S.S.R.* **37**, 32.
Vitkevich, V. V.: 1960b, *A. J. U.S.S.R.* **37**, 961.
Vitkevich, V. V.: 1968, *Astron. Tsirkul. U.S.S.R.* **477**.
Vitkevich, V. V. and Vlasov, V. I.: 1966, *Astron. Tsirkul. U.S.S.R.* **396**.
Vitkevich, V. V. and Vlasov, V. I.: 1968, *Dokl. Akad. Nauk U.S.S.R.* **181**, 572.
Vitkevich, V. V. and Vlasov, V. I.: 1969, *A. J. U.S.S.R.* **46**, 851.

GALACTIC COSMIC RAY MODULATION BY THE
INTERPLANETARY MEDIUM
(INCLUDING THE PROBLEM OF THE OUTER BOUNDARY)

L. I. DORMAN
IZMIRAN, Moscow, U.S.S.R.

Abstract. This paper reviews the present status of the study of the galactic cosmic ray modulation effects, such as 11-yr variations, radial gradients, yearly variations, transverse gradients, 27-day variations, sporadic variations, Forbush decreases, cosmic-ray intensity increases before a Forbush decrease, solar anisotropy of the galactic cosmic rays and its time variations, microvariations, etc. Theoretical concepts which attempt to explain the main features of the observed events are discussed. Information about the properties of the solar wind, its magnetic field, and propagation of disturbances derived from the data on the cosmic-ray modulation effects are discussed. The 11-yr variations and their connection with various solar activity indices, in particular with the HL-index which takes account of the distribution of solar activity in heliocentric latitude, are discussed very thoroughly. The properties of the solar wind at the boundary with the interstellar medium are estimated. The turbulent transition zone of the subsonic solar wind and its possible effect on the galactic cosmic rays are studied.

1. General Picture of Galactic Cosmic-Ray Modulation in Interplanetary Space

This picture is based on Parker's (1963) concept of the solar wind, the regular, and irregular magnetic field moving with the solar wind, magnetic disturbances propagating in interplanetary space and the interaction of fast charged particles (protons, various nuclei, electrons) or cosmic rays with these magnetic fields moving radially from the Sun (Parker, 1958). As a result of such an interaction cosmic rays are carried convectively away from the solar system and their density is decreased, but a diffusion of the cosmic-ray flux from the interstellar to the interplanetary medium also appears. Let u be the solar wind (bulk) velocity, r distance from the Sun, $R = pc/Ze$ the particle rigidity (in which p, c, Z and e have their conventional meanings of momentum, the velocity of light, the atomic number, and electron charge respectively), t the time, and $n(r, Ze, R, t)$ the distribution function for cosmic-ray density in interplanetary space. The convective flux directed from the Sun will be nu. One can easily see that, because of this flux, the volume would be very rapidly emptied of the cosmic-ray 'gas'. In fact, in the observations near the Earth's orbit, at $r = r_{\oplus} = 1 \text{AU} = 1.5 \times 10^{13}$ cm and $u = 3 \times 10^7$ cm/sec, we have $\mathrm{d}/\mathrm{d}t((4\pi/3)nr_{\oplus}^3) = -4\pi r_{\oplus}^2 nu$, whence $n = n_0 e^{-t/\tau}$ in which the time constant $\tau = r_{\oplus}/3u = 1.5 \times 10^5$ sec, or less than two days. That such a quick emptying does not take place is due to the presence of the diffusion flux $-\kappa \nabla n$, where κ is the diffusion coefficient. Within a fairly short time a quasistationary distribution of the cosmic-ray density is established by the balancing of the convection and diffusion fluxes $nu = \kappa \nabla n$. From this we can obtain, for the spherically symmetrical model:

$$n(r, Ze, R, t) = n_0(Ze, R) \exp\left[-\int_r^\infty \frac{3u(r, t)\,\mathrm{d}r}{v(Ze, R)\,\Lambda(n, R, t)} \right], \tag{1}$$

Dyer (ed.), Solar Terrestrial Physics/1970: Part II, 67–91. All Rights Reserved.

where $n_0(Ze, R)$ is the non-modulated spectrum of the primary cosmic-ray density in interstellar space, $v(Ze, R) = CR[R^2 + (Am_0c^2/Ze)^2]^{-1}$ is the particle velocity, and $\Lambda(r, R, t)$ is the mean free scattering path (here it has been included in $\kappa = v\Lambda/3$). As the solar activity increases, u increases and κ decreases, which results in an enhancement of the modulation and a decrease in the galactic cosmic-ray intensity. This is the well known 11-yr cosmic-ray variation the amplitude of which, A_{11}, is substantially dependent on particle rigidity R. In the range $R \approx 20 - 30$ GV (measured with a neutron monitor at the equator) $A_{11} \approx 5$–6%, at $R \approx 7$–10 GV (with neutron monitors at medium and high latitudes) $A_{11} \approx 20\%$; at $R \approx 1$–3 GV (balloon observations at high latitudes) the cosmic ray intensity is decreased by a factor of almost two from minimum to maximum solar activity. Such a strong dependence of A_{11} on R in the relativistic domain is determined by the dependence of Λ on R (Λ increases with increasing R); in the non-relativistic domain it is determined mainly by the rapid decrease in v with decreasing R. It follows from Equation (1) in particular, that modulation will distort the primary differential rigidity spectrum, with the strongest distortion being in the low-rigidity range. In the relativistic domain where $v \approx c$ the modulation is a function of particle rigidity only, so that the relative chemical composition of cosmic rays (i.e., according to Ze) at identical rigidities inside the solar system at any modulation depth should be the same as in the interstellar medium. Substantial differences should be expected only in the non-relativistic domain for different values of the ratio of atomic mass to atomic number A/Z; in particular, if A/Z increases at a given rigidity R, then v decreases, which should result in a relative increase in the modulation depth and in a depletion of such particles in the charge composition inside the solar system as compared with the composition in the interstellar medium.

It should be noted that the first simple model proposed by Parker made it possible at once to understand qualitatively the nature of the 11-yr cosmic-ray variation; but in order to explain the dependence of A_{11} on particle rigidity an additional assumption had to be introduced concerning the presence of a broad spectrum of magnetic inhomogeneities in interplanetary space (Dorman, 1960, 1963a, b). This assumption was further confirmed by direct measurements in interplanetary space (Heppner *et al.*, 1962; Ness, 1966; Coleman, 1967). A strict theory establishing the connection between Λ and the spectrum of magnetic inhomogeneities was developed by Dolginov and Toptigin (1968), Tverskoy (1968), Jokipii (1966), Jokipii and Coleman (1968), and Roelof (1966). Later on, Parker's idea was substantially developed in a number of works which provided for the inclusion of many of the real properties of the solar wind and its interaction with the galactic cosmic rays: for instance, (1) cosmic-ray deceleration due to the diffusion of cosmic rays in the expanding solar wind, a phenomenon that results in an increase of the amplitude of the modulation, so the exponent in Equation (1) is increased by a factor of 1.5 (Singer, 1958; Dorman, 1963c; Parker, 1963, 1965; Krymsky, 1966); (2) replacement of the diffusion coefficient by a diffusion tensor which takes account of the anisotropic propagation of cosmic rays in interplanetary space, due to the presence of the regular quasispiral field (in addition to the magnetic inhomogeneities); this permitted, in particular, the nature of the solar-diurnal cosmic-ray variation to be understood (Krymsky, 1964; Parker, 1965; Gleeson and Axford, 1967, 1968; Axford, 1965; and Dorman, 1965, 1966, 1967); (3) inclusion of the changes of solar wind with heliocentric longitude, which led to an understanding of the 27-day cosmic-ray variation (Dorman, 1961; Belov and Dorman, 1969, 1971; Dorman and Kobylinsky, 1968); (4) the effect of moving 'semitransparent magnetic pistons' (shock waves) which explains Forbush-decreases (Parker, 1963; Dorman, 1963b; Blokh et al., 1964), and the cosmic-ray intensity increases before the magnetic storm commencement (Blokh et al., 1959, 1960, 1961; Dorman et al., 1969a, b, 1970); (5) the time delay of certain electromagnetic phenomena in inter-planetary space with respect to the corresponding events or processes on the Sun, which is especially important for the investigation of large modulation regions (Dorman and Dorman, 1967a, b, 1968a); (6) attempts to include the inverse effect of cosmic rays on solar wind (Axford and Newman, 1965; Dorman and Dorman, 1968b, c, d, 1970; Sousk and Lenchek, 1969).

The main effect of galactic cosmic-ray modulation is the 11-yr variation against the background of which the 27-day variations develop. It appears that these variations are due to two causes: (1) asymmetry in the solar wind due to long-lived asymmetry of solar activity (i.e. lasting for several solar rotations; (2) 27-day recurrence of the Forbush decreases, also due to long-lived active regions. After powerful chromo-spheric flares, strong shock waves propagate in the interplanetary medium, the pecu-liar moving semi-transparent magnetic 'pistons' creating a rarefaction in the cosmic-ray gas behind the 'piston' (corresponding to a Forbush decrease) and a condensation in front of the 'piston' (corresponding to the short increase in intensity before the sudden-commencement magnetic storms). The quasispiral regular magnetic field is of considerable importance in the interplanetary region close to the Sun (up to distances of several AU from the Sun), and the diffusion flux is directed, not radially, but mainly along the field. In this case, under quasistationary conditions, the radial component of the diffusion flux is balanced by the convection flux (for a given distribution of the galactic cosmic-ray density in interplanetary space determined by expression (1)), whereas the tangential component of this diffusion flux is not compensated. This causes the solar-diurnal cosmic-ray variation, the peak of which comes at 18h LT, when observed outside the magnetosphere. Distortion of the particle trajectories inside the geomagnetic field shifts the time of the peak by several hours (to noon). Of course, at particular times the convection flux and radial component of the diffusion flux are *not* balanced, and the resulting transient solar anisotropy of the galactic cosmic rays may then differ substantially from the stationary anisotropy (i.e. averaged over a long period). Examples of such solar anisotropies, e.g., abrupt changes in the solar-diurnal variation during Forbush decreases, or when going from one solar wind sector to another, or the 27-day and 11-day variation, etc., are well known.

Of course, the flux of magnetic inhomogeneities is not regular over heliocentric latitude. It should be expected that the most marked modulation would occur in the low-latitude region where solar activity is concentrated. This should result in a non-uniform latitudinal distribution of the cosmic-ray density, and thus in the appearance of transverse gradients. The cosmic rays would be 'sucked in' from the north and south and then carried off in the direction away from the Sun, i.e. the so-called 12 hr solar anisotropy of cosmic rays would appear.

The modulation of cosmic rays is believed to cease at some distance r_0 from the Sun. (This distance r_0 could be a function of the rigidity.) There are at least two reasons for believing this: (1) the magnetic fields would eventually become so weak that the particle scattering would become ineffective (i.e., in Equation (1)$\Lambda \rightarrow \infty$); (2) because of interaction of the solar wind with low-energy galactic cosmic rays, with the magnetic field of the surrounding galactic (interstellar) medium, or as a result of charge exchange, the solar wind velocity **u** should decrease substantially, so that according to Equation (1) the contribution of the region with $r > r_0$ would become insignificantly small.

It is possible that there exists a transition region where the solar wind becomes subsonic, with intricately tangled magnetic fields at the boundary with the interstellar gas. Such as region should affect the low-energy particles especially strongly and cause a delay of the recovery of their intensity after a period of high solar activity.

These are, in brief, the physical foundations of the general picture of the galactic cosmic-ray modulation in interplanetary space. We cannot discuss here in detail all the various types of the modulation effects mentioned above (for a treatment of some of these, the reader is referred to other reviews, e.g. Parker, 1963; Dorman, 1963a, b; Gleeson, 1968; and Krymsky, 1969). Here we shall dwell only on those problems which are, in the opinion of the author, either the most critical or the most interesting, or which are to a considerable degree unsolved. We hope that this discussion will attract the attention of investigators and will be helpful in solving the problems.

2. The 11-yr Cosmic-Ray Variations

2.1. VARIOUS ASPECTS AND METHODS OF STUDY

The 11-yr cosmic-ray variations reflect most completely the substantial effect of solar activity on galactic cosmic rays. A thorough study of these variations is of great interest from several viewpoints. First, the data on the 11-yr variations, which are the integral effect of the solar wind interaction with cosmic rays entering interplanetary space from the Galaxy, contain valuable information on the properties of the solar wind itself (its velocity, structure, magnetic field strength, dimensions, distribution in space, etc.). Second, all modulation effects of cosmic rays are closely connected with each other; thus, an understanding of the nature of the 11-yr variations is important for an understanding of many other effects, such as the 27-day variations, solar-diurnal variations, etc. It is also important for an understanding of the process of the solar cosmic-ray propagation in interplanetary space. The fact is that the electromagnetic characteristics of interplanetary medium which can be derived from a study of the 11-yr cosmic-ray variations are also a determinant for the 27-day and solar-diurnal variations, Forbush effects during magnetic storms, and for the modification of these effects with changes in the level of solar activity. Third, the study of the

11-yr variations makes it possible to determine how the original differential energy spectrum of galactic cosmic rays is distorted by the interplanetary medium and to reconstruct the undistorted spectrum for the primary cosmic rays with various A/Z ratios outside the solar system. This in turn is of paramount importance for the solution of a number of problems concerning the origin of cosmic rays. Fourth, the study of the 11-yr variation may help to estimate the effect of masking and distortion of the original anisotropy (if any) of the galactic cosmic rays that are introduced if the measurements are made within the solar system. Fifth, a thorough study of the 11-yr variations makes it possible to develop a more comprehensive theory of the cosmic-ray interaction with the solar wind and to show, in particular, that the conventional interaction model based on the linear approximation (in which the solar wind is considered as given and only its unilateral effect on cosmic rays is studied) are too rough, at least for great distances from the Sun. For this region, the non-linear problem, including the inverse effect of cosmic rays on the solar wind, must be solved. Finally, it should be noted that at present the 11-yr variation seems to be the only source of information on the transition region or boundary between the solar wind and interstellar medium.

On their way to Earth, primary cosmic rays pass through regions that produce intensity variations of quite different kinds. For simplicity, we can identify three regions: (1) the Earth's atmosphere, (2) the magnetosphere, and (3) the interplanetary medium, responsible for major changes in flux, spectrum, and nuclear composition of the primary cosmic rays. Measurements made at the bottom of the Earth's atmosphere are subject to all three influences, measurements made above the atmosphere are subject to the last two, and those outside the magnetosphere in interplanetary space are subject only to the last. However, as shown by Dorman (1970), only the last region and the effects it produces are of particular importance for the 11-yr cycle: the effects of the first two can undoubtedly be neglected if sliding 12-months mean of the observational data obtained in the first two regions are used for the analysis of the 11-yr variation.

The 11-yr cosmic ray variations may be studied in three ways (Dorman, 1970):

(1) On the basis of data on the variation, with solar activity, of the geographic distribution of the intensities of various secondary components. Such information is fairly regularly obtained from various special expeditionary studies. This method makes it possible to find, with high accuracy, the 11-yr variation in the primary spectrum in the rigidity range from 15–18 GV to 1–2 GV from ground-based observations, and to extend the spectrum down to energies of 100–200 MeV with stratospheric studies.

(2) On the basis of continuous observations made by the world-wide network of fixed stations. These data provide valuable information on the relation between the 11-yr variations at various energy ranges and some of the characteristics of solar activity.

(3) On the basis of measurements of the energy spectrum and chemical composition of galactic cosmic rays (including electrons) on balloons, satellites, and space probes.

These data are of exclusive importance for the study of the 11-yr variations in the range of very low energies. Simultaneous measurements at various distances from the Sun and from the ecliptic plane provide direct information on the radial, transverse, and tangential gradient of cosmic rays in interplanetary space.

2.2. The 11-yr variation spectrum and its change from year to year according to the data on latitude effects

It is easy to show (Dorman and Dorman, 1965a) that, if the cutoff rigidity R_c and the integral multiplicity $m^i(R, h_0)$ are constant over the 11-yr solar activity cycle, the primary spectrum of the 11-yr variation can be directly determined from the results of measurements of the latitude effects of various cosmic-ray components:

$$\frac{D(R_c, t)}{D_{\min}(R_c)} = \frac{\partial N^i(R_c, h_0, t)/\partial R_c}{\partial N^i_{\min}(R_c, h_0)/\partial R_c};$$

$$\frac{\delta D(R_c, t)}{D_{\min}(R_c)} = \frac{D(R_c, t) - D_{\min}(R_c)}{D_{\min}(R_c)} = \frac{\partial N^i(R_c, h_0, t)/\partial R_c}{\partial N^i_{\min}(R_c, h_0)/\partial R_c} - 1$$

(2)

Here D is the cosmic-ray intensity and N^i the intensity of the ith secondary component; the subscript 'min' signifies that the value of the variable is that prevailing at solar minimum.

In this case, errors connected with inaccurate values of the coupling coefficients and integral multiplicities in the generation of secondary components are avoided.

Expression (2) was used by Dorman (1971) to analyze the observational data for the years 1954–1962 on the neutron-component latitude effects obtained by Japanese investigators on board the ships 'Labrador', 'Atka', 'Arneb' and 'Soya', which cruised from Japan to Antarctica and back (Kodama, 1968). The primary spectra $\delta D(R, t)/D_{\min}(R)$ were derived, where $D_{\min}(R)$ is the primary cosmic ray spectrum in 1954. The results obtained may be summarized as follows (see Figure 1):

(a) In the low-rigidity range ($R \lesssim 3$ GV) the form of the 11-yr variation spectrum $\delta D/D_{\min}$ changes considerably with the solar cycle. If the spectrum is represented in exponential form $\delta D/D_{\min} \propto R^{-\gamma}$, the exponent varies from $\gamma \approx 1$ in the period of low solar activity (1955–1956) to $\gamma \approx 0.5$ during high solar activity (1957–1959). As the solar activity decreases, γ does not immediately increase, as would be expected if there is a simple connection between γ and solar activity; instead, γ continues to decrease down to the value $\gamma \approx 0.2$ (1960–1961). Not until 1962 does the value of γ increase somewhat, up to $\gamma \approx 0.4$–0.5. Thus, in the low-rigidity range, there is a 2–3 yr delay in the variation of the exponent γ with respect to the level of solar activity after the maximum.

(b) In the medium-rigidity range (3–8 GV) the exponent varies only slightly with the solar activity cycle, that is, within the range $\gamma \approx 1.6$–1.7.

(c) In the high-rigidity range ($R \gtrsim 8$ GV), during the phase of low solar activity, the relative changes in the intensity barely exceed the fluctuations, so that it is rather difficult to draw a definite conclusion about the character of the spectrum in this region.

During the high solar activity phase $\gamma \approx 1.7$–1.6, i.e. within the same limits as for $3\,\mathrm{GV} \lesssim R \lesssim 8\,\mathrm{GV}$.

(d) There is a clear tendency for γ to increase with rigidity, from 0.2–0.5 at $R \gtrsim 3\,\mathrm{GV}$ to 1.6–1.7 at $R \gtrsim 3\,\mathrm{GV}$ during the high solar activity phase. Important conclusions can be drawn about the nature of the dependence of the mean free scattering path Λ on particle rigidity R. For example, if it is assumed to a first approximation that the size (effective radius) of the modulation region r_0 is independent of R, then, since for small modulation amplitudes (e.g., neutron component observations) $(\delta D(R,\,t)/D_{\min}(R)) \propto$ $\propto (ur_0/v\Lambda)$, where u is the solar wind velocity, and v is the particle velocity, we obtain

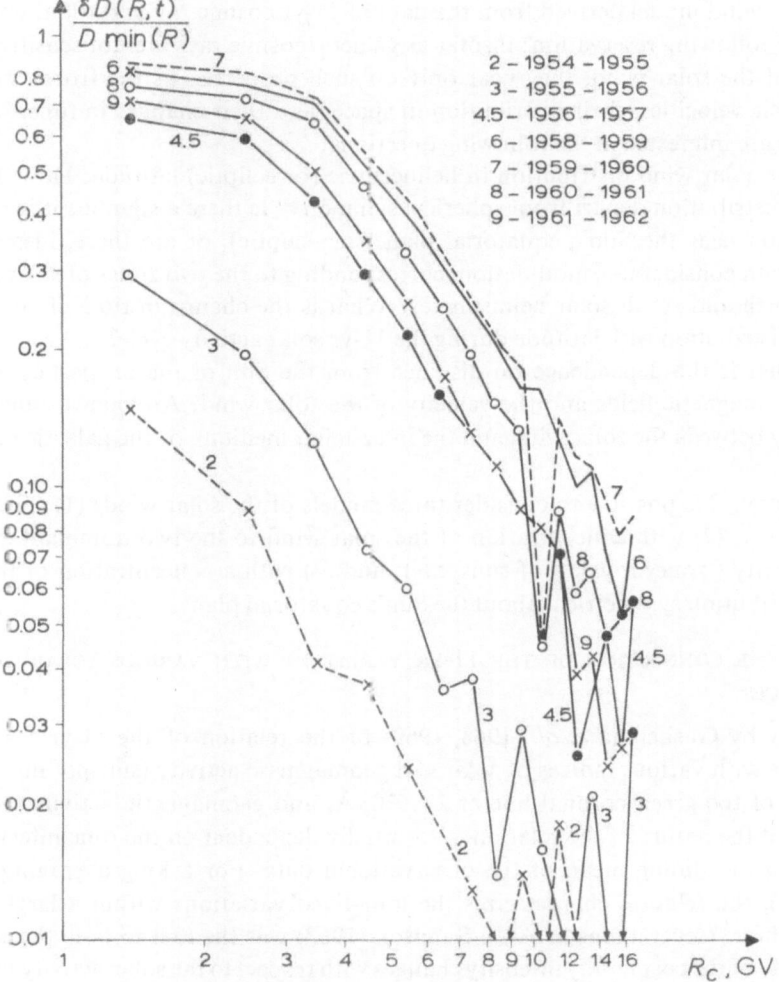

Fig. 1. Determination of $\delta D(R,\,t)/D_{\min}(R)$ on the basis of the measurements of the neutron-component latitude effects for eight Japanese expeditions.

$\Lambda \propto R^{\gamma}$. (In this case $\gamma = 1.6$–1.7 at $R > 3$ GV.) However, if allowance for the possible dependence of r_0 on R is included (as shown by Dorman and Dorman, 1967d), r_0 would be expected to decrease somewhat with increasing R, for example as $r_0 \propto R^{-\beta}$, $(\beta > 0)$, in the high rigidity range, in which case $\Lambda \propto R^{\gamma + \beta}$.

A similar study of the 11-yr variation of the spectrum by Dorman (1970), based on the results of ground-based and stratospheric measurements at various latitudes during Soviet expeditions in 1964, 1965, 1966, and 1968, shows that the main features of the spectrum found for solar activity cycle No. 19 (1954–1964) are maintained in cycle No. 20 (beginning in 1964).

2.3. THREE POSSIBLE MODELS OF SOLAR WIND

In a solar wind model derived from the data of 11-yr cosmic ray variation, we should make the following reservation: insofar as galactic cosmic rays are not sensitive to the density of the solar wind, they bear only on such properties as the frozen magnetic fields, their velocities, their distribution in space, and their changes in time. In particular, we are interested in the following questions:

(1) The solar wind distribution in heliocentric (or ecliptic) latitude: How does the latitude distribution depart from spherical symmetry? Is there a significant increase of modulation near the Sun's equatorial plane (or ecliptic), or are there, instead, two regions with considerable modulation, corresponding to the two zones of solar activity in the north and south solar hemispheres? What is the change in time of cosmic-ray density distribution with latitude during the 11-yr solar activity cycle?

(2) What is the dependence on distance from the Sun of the properties of interplanetary magnetic fields and the velocity of the solar wind? At what distance is the boundary between the solar wind and the interstellar medium, or the galactic magnetic field?

In general, it is possible to consider three models of the solar wind: (1) spherically-symmetrical; (2) with concentration of the solar wind to the two dominant zones of solar activity ('zones royales' of sunspots); and (3) with a concentration of the solar wind distribution symmetrical about the Sun's equatorial plane.

2.4. ON THE CONNECTION OF THE 11-YR VARIATION WITH VARIOUS SOLAR ACTIVITY INDICES

The study by Gushchina *et al.* (1968, 1969) of the relation of the 11-yr cosmic ray variations with various indices of solar and geomagnetic activity (sunspot number W, intensity of the green coronal line at $\lambda = 5303$ Å, and geomagnetic activity index A_p) shows that the nature of the relation is essentially dependent on the time interval used for taking the sliding mean of the observational data. For a long averaging period (0.5–1 yr), the relation characterizes the long-lived variations within a large modulation volume (several tens of AU). Simpson (1963) was the first to note the considerable delay of the cosmic-ray intensity changes with respect to the solar activity changes. Figure 2 shows, in particular, that the delay Δt_1 accompanying an increase in solar activity is considerably smaller than Δt_2, accompanying the activity decrease – in fact, by a factor of about 2.6 in this case. This asymmetry increases with decreasing particle energy. The time of the delays is of the order of a year.

We believe that a delay of such a duration is due to two causes: (1) a large volume for the modulation region with radius $r_0 \approx ut \approx (3 \times 10^7 \text{ cm/sec}) \times (3 \times 10^7 \text{ sec}) \approx \approx 10^{15} \text{ cm} \approx 100 \text{ AU}$; (2) the presence of a transition zone of subsonic solar wind at the boundary with the interstellar medium, which results in an increased delay in the

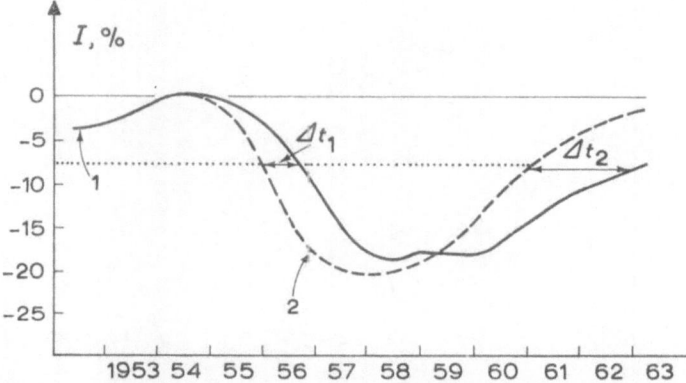

Fig. 2. Changes during 1952–63 in the cosmic ray neutron-component intensity at a station with $R_c = 1.5$ GV (solid curve 1) and the changes in the sunspot number W (dashed curve 2), both in percent deviation relative to the 1954 level.

recovery for low-energy fluxes after solar maximum. A third possible cause is discussed in the literature and might be added to the first two, namely: (3) the delay is due to the shift of the latitude that zones of solar activity from low to higher heliocentric latitudes as the solar cycle develops. In fact, this explanation also gives a 1–2 yr delay (Kudo and Wada, 1968; Stozhkov and Charakhchian, 1969). But the last viewpoint is subject to the following objections:

(1) If the delay is due only to latitude shift, it should be the same for all energies, which sharply contradicts the observations.

(2) If the latitude shift were the main cause of the delay, the cosmic-ray distribution in the interplanetary medium should depart substantially from the spherical symmetry, i.e. large transverse (latitudinal) intensity gradients should appear. However, the analysis of the yearly variations carried out by Dorman et al. (1967a, b) shows that the transverse gradients are in any case an order of magnitude smaller than these expected according to this mechanism.

(3) Dorman and Kobylinsky (1968) calculated the numerical solution of the equation of the cosmic-ray propagation in interplanetary space for the case when the solar wind velocity u and the coefficient of diffusion κ depend on the heliocentric latitude θ in such a manner that the modulation parameter u/κ is decreased by a factor of 4 from the equator to the pole. Figure 3 shows that the expected difference in the intensity at $\theta = 0$ and $\theta = \pi/2$ for $r \lesssim 0.1 r_0$ is insignificantly small; i.e. even with such a strong

dependence of u/κ on θ, the result is almost the same as for the spherically symmetrical model. It follows that when studying the 11-yr variation we may use the spherically symmetrical model for modulation, and use the same solar activity index over the whole disc.

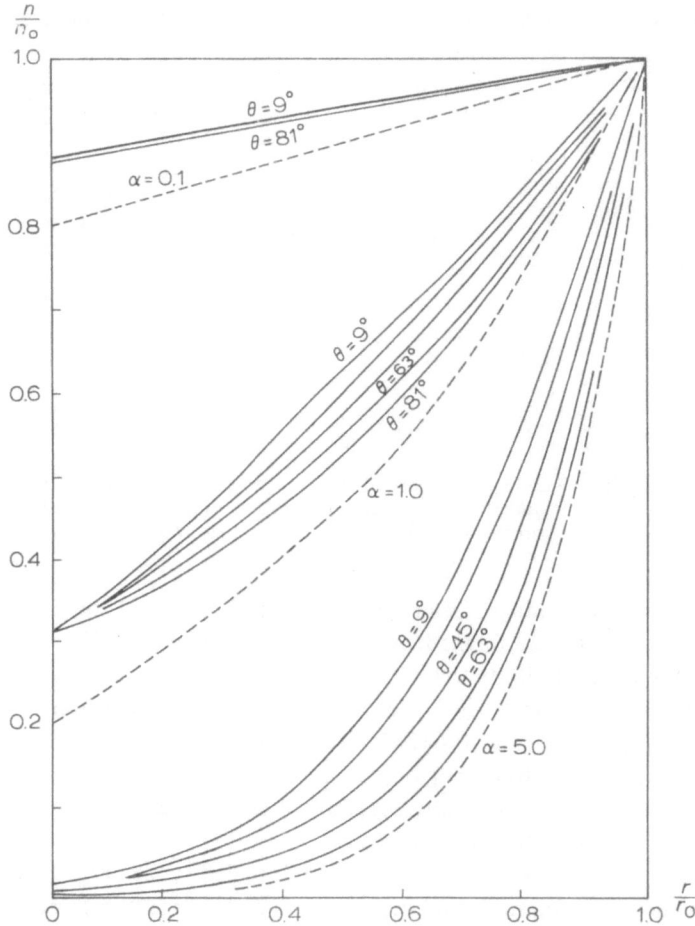

Fig. 3.   Dependence of the modulation depth n/n_0 on r/r_0 and θ for $\alpha = ur_0/\kappa = 0.1$, 1.0, and 5.0 (solid curves), according to Dorman and Kobylinsky (1968). Shown with dashed lines are the solutions for the spherically symmetrical model with $\kappa = \kappa_0$ and $u = u_0$, according to Dorman (1967a).

2.5. POSSIBILITY OF USING A STATIONARY MODEL

It appears that a quasistationary model may also be used with sufficient accuracy. Let us take the solution of Parker's equation for the spherically symmetrical model. From the 11-yr variation in spectra at $R \approx 3$ GV presented above, we have $(3ur_0/v\Lambda) = 0.9$. Assume that $r_0 = 10^{15}$ cm; from this $\Lambda = 3ur_0/0.9v = 3 \times 10^{12}$ cm. Then one estimates the time of cosmic-ray diffusion $t_{dif} = (r_0^2/2v\Lambda) = (10^{30}/2 \times 3 \times 10^{10} \times 3 \times 10^{12}) \approx$ $\approx 5 \times 10^6$ sec≈ 2 months. On the other hand, the characteristic time of the cosmic-ray intensity change, $t_{change} = 11$ years$/2\pi \approx 2$ yr; i.e., $t_{change} \gg t_{dif}$. Hence the stationary equation may be used and the time t may be considered as a parameter. It will be noted that if r_0 is an order of magnitude smaller, say $r_0 = 10^{14}$ cm, then $\Lambda (3 \text{ GV}) \approx 3 \times 10^{11}$ cm; but this is almost an order of magnitude smaller than the estimate from solar cosmic-ray data (Dorman and Miroshnichenko, 1968).

2.6. Inclusion of the Delay in the Spherically Symmetrical Model

With a large modulation volume, one must take into account the delay of electromagnetic conditions in interplanetary space with respect to the events on the Sun which caused them. Thus, if the conditions are of the type transported by the solar wind with velocity u, then conditions in interplanetary space at a moment of time t and a distance r from the Sun were determined by events on the Sun occurring at an earlier time $t - (r/u)$. This circumstance can be included in the spherically symmetrical, quasistationary model of modulations. Equating the diffusion flux $\kappa\nabla n$ to the convection flux nu (as before) we obtain

$$\frac{v(R)\,\Lambda(r, t, R)}{3}\frac{\partial n(r, t, R)}{\partial r} = u_{dr}(r, t)\,n(r, t, R) \tag{3a}$$

in which the solar wind velocity u is replaced by u_{dr}, the radial drift velocity of cosmic-ray particles carried along by the regular expanding spiral field and the accompanying inhomogeneities. In general $u_{dr} \lesssim u$. The solution is

$$\frac{n(r, t, R)}{n_0(R)} = \exp\left[-\frac{3}{v(R)}\int_r^{r_0}\frac{u_{dr}(r, t)}{\Lambda(r, t, R)}\,dr\right], \tag{3b}$$

which is the counterpart of Equation (1), with the boundary condition $n(r_0, t, R) = n_0(R)$ introduced.

Now, let W be the solar activity index (actually the Wolf sunspot number), N the number of magnetic inhomogeneities per unit volume, d the characteristic distance between inhomogeneities, and l their characteristic size. Then $N = d^{-3}$. Furthermore N increases with solar activity; we shall assume the simple proportionality $N(t) \propto W(t - r/u)$, in which r/u is the time delay at distance r, as noted above.

Let us consider two extreme cases:

(1) High solar activity, in which the volume is effectively filled with inhomogeneities, so that $d \approx l$. Then, the mean free path for scattering is given by:

$$\Lambda(r, t, R) \approx d^3/l^2 \approx d = N^{-1/3}(t) \propto W^{-1/3}(t - r/u). \tag{4}$$

(2) Low solar activity, for which the inhomogeneities are widely separated, so that $d \gg l$. Then, taking l as constant,

$$\Lambda(r, t, R) \approx d^3/l^2 \propto d^3 = N^{-1}(t) \propto W^{-1}(t - r/u). \tag{5}$$

Then, for the general case, we can write

$$A(r, t, R) = A_{max}(r, R) \left[\frac{W(t - r/u)}{W_{max}} \right]^{\alpha},$$ (6)

where the subscript max indicates values for solar maximum, and the exponent α can take various values, e.g.$-\frac{1}{3}$ in Equation (4), and -1 in Equation (5).

If we allow for anti-Fermi deceleration, the coefficient $-3/v(R)$ in the exponent on the right-hand side of Equation (3) should be replaced by $-(\gamma+2)/v(R)$, in which γ is the spectral index for primary cosmic rays. It has the value ≈ 2.5, so

$$\gamma + 2 \approx 4.5.$$ (7)

If $A_{max}(r, R) = A_{max}(R)$, and u_{dr} is independent of r, then with the substitution $\chi = r/r_0$

$$\frac{n(r_{\oplus}, t, R)}{n_0(R)} = \exp\left\{ -\frac{4.5 u_{dr} r_0}{v(R) A_{max}(R)} \int\limits_{r_{\oplus}/r_0}^{1} \left[W\left(t - \frac{r_0}{u}\chi\right) \middle/ W_{max} \right]^{-\alpha} d\chi \right\}.$$ (8)

If $A_{max}(r, R) = A_{\oplus}(R)r/r_{\oplus}$, then

$$\frac{n(r_{\oplus}, t, R)}{n_0(R)} = \exp\left\{ -\frac{4.5 u_{dr} r_{\oplus}}{v(R) A_{\oplus}(R)} \int\limits_{r_{\oplus}/r_0}^{1} \left[W\left(t - \frac{r_0}{u}\chi\right) \middle/ W_{max} \right] \frac{d\chi}{\chi} \right\}.$$ (9)

Let us examine an idealized model in which $(W/W_{max})^{-\alpha} = \frac{1}{2}(1 + \cos\omega t)$, in which ω is the 'angular frequency of the solar cycle', i.e. 2π rad/11 yr or 1.9×10^{-8} rad/sec. Substituting into Equation (8), integrating, and restricting ourselves to the case of weak modulation $\Delta n/n_0$, so that the approximation $e^{-y} - 1 \approx -y$ applies, we have:

$$\frac{\Delta n}{n_0} \approx -\frac{4.5 u_{dr} r_0}{2v(R) A_{max}(R)} \left[1 - \frac{r_{\oplus}}{r_0} + \frac{2u}{r_0\omega} \sin\frac{\omega}{2u}(r_0 - r_{\oplus})\cos\omega\left(t - \frac{r_0 + r_{\oplus}}{2u}\right) \right].$$ (10)

From this we can conclude that: (1) if the frequency of the solar cycle is ω, the frequency of the cosmic ray intensity variations is also ω; (2) the delay time $t_{delay} = (r_0 + r_{\oplus})/2u$; (3) the amplitude of the modulation is proportional to $(2u/r_0\omega)$ sin $(\omega[r_0 - r_{\oplus}]/2u) \approx (2u/r\omega)$ sin $(r_0\omega/2u)$, $(r_0 \gg r_{\oplus})$.

The results for two cases are of special interest:

Case (a): For slow variations (e.g. the 11-yr variation, ω small), $r_0\omega/2u \gtrsim 1$, then the coefficient of the cosine term in Equation (10) is ≈ 1 and $\Delta n/n_0 \propto 1 \times \cos\omega[t-(r_0+r_{\oplus})/2u]$ approximately. For the 11-yr variation, $r_0\omega/2u \approx 10^{15}$ cm $\times 1.9 \times 10^{-8}$ rad sec^{-1}/$2 \times 3 \times 10^7$ cm sec$^{-1} \approx \frac{1}{3}$.

Case (b): With more rapid variations, $r_0\omega/2u \gg 1$, then $2u/r_0\omega$ sin $(\omega[r_0-r_{\oplus}])/2u < 2u/r_0\omega \ll 1$, and the amplitude diminishes. For example, if $T = 2\pi/\omega \approx 0.5$ yr, then $2u/r_0\omega \approx 0.1$, and the amplitude of $\Delta n/n_0$ is $\approx 2u\, r_0^*/v A$, in which r_0^* is the radius of a smaller modulation volume corresponding to the more rapid fluctuations; in this case $r_0^* \approx 0.1 r_0$.

It follows from the idealized model just examined, that the study of the whole modulation volume requires the use of data averaged over a long period (0.5–1 yr). When averaging over one month the main contribution to the correlation is from more rapid changes, for which regions of radius $r_0^* \ll r_0$ are effective. For example, if the mean daily values are used and the delay is 5–6 days, then $r_0^* \approx 1$–2 AU.

We shall now include the real changes in solar activity. This may be done by calculating the integrals

$$b = \int_{r_0^*/r_0}^{1} \left[\frac{W\left(t - \dfrac{r_0}{u}\chi\right)}{W_{max}} \right]^{-\alpha} d\chi; \qquad d = \int_{r_0^*/r_0}^{1} \left[\frac{W\left(t - \dfrac{r_0}{u}\chi\right)}{W_{max}} \right]^{-\alpha} \frac{d\chi}{\chi} \quad (11)$$

from solar activity data (Dorman and Dorman, 1967a, b, c).

These integrals depend only on the parameter r_0/u. An unknown adjustment or scale conversion factor before b and d may be determined from a comparison of the theory with experiment. Figure 4 shows such a comparison with data from the Chicago station for the case $\Lambda_{max}(r, R) = \Lambda_{max}(R)$. The calculations and the observational data have been tied together at the minimum and maximum of cosmic-ray intensity. It can be seen that, with $r_0/u = 10$ months, the hysteresis is much smaller than that required to explain the observations. With $r_0/u = 20$ months, experimental and theoretical results are in good agreement. In both cases $\alpha = -\frac{1}{3}$. The scale conversion factor

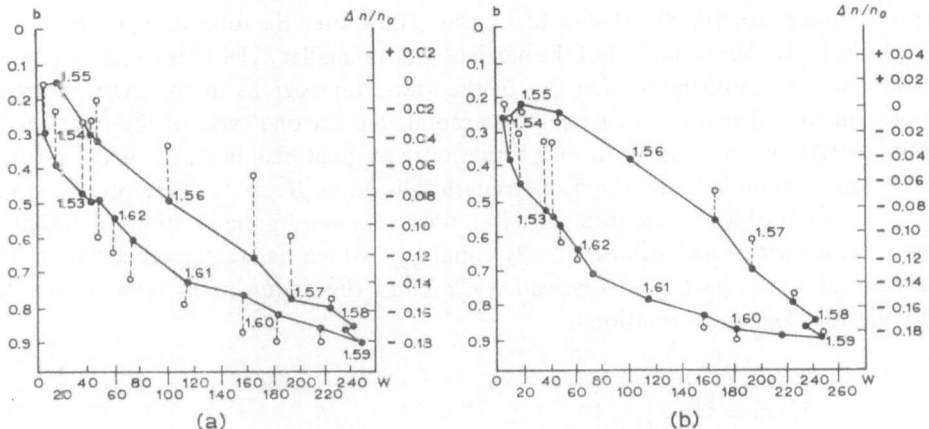

Fig. 4. Predicted changes in the integral b (dots connected with solid lines; see Equation (11)) as a function of the solar activity index W smoothed with a period of 12 months, and a comparison with the observational data $\Delta n/n_0$ of the neutron monitor at Chicago (represented by circles connected to the corresponding theoretical values by dashed vertical lines). The calculated and observed points are at six-month intervals for the 10 years from January 1953 to January 1963; the points for each January are marked 1.53, 1.54, etc. beginning near $W \approx 40$, and follow the hysteresis loop in a clockwise direction: (a) The values of b for $r_0/u = 10$ months and $\alpha = -\frac{1}{3}$; (b) for $r_0/u = 20$ months and $\alpha = -\frac{1}{3}$.

from the right to the left scale for Figure 4 is 0.275, whence $\Lambda_{max}(7\text{--}10\ \text{GV}) \approx 1$ AU and $r_0 = 100$ AU for $u = 3 \times 10^7$ cm/sec. If the solar wind were entirely absent, the intensity would be 5% higher than that in 1954, i.e. some modulation is maintained, even at solar activity minimum.

In the foregoing case Λ increases considerably with decreasing solar activity. Let us now consider the case in which Λ increases in proportion to r. In this case, theory and observation disagree, even with $r_0/u = 80$ months, i.e. $r_0 = 400$ AU. But this high estimate contradicts the estimates of r_0 obtained from equating the solar wind pressure, on the one hand, to the pressure of the galactic magnetic field and low-energy galactic cosmic rays on the other:

$$r_0/r_\oplus = \min\left\{4 \times 10^{-12}\frac{u\sqrt{n_\oplus}}{H_{\text{gal}}};\ \ 10^{-6}\ u\ \sqrt{\frac{n_\oplus}{W_{CR}}}\right\}. \tag{12}$$

Here $n_\oplus$ is the solar wind concentration near the Earth's orbit, H_{gal} is the strength of the magnetic field in the galactic surroundings or spiral arm and W_{CR} is the cosmic-ray energy density in interstellar space (in eV/cm^3). With $u = 4 \times 10^7$ cm/sec, $n_\oplus = 5\text{cm}^{-3}$, $H_{\text{gal}} = 3 \times 10^{-6}$ G, and $W_{CR} \approx 1\text{--}2$ eV cm^{-3}, this gives $r_0 \leqslant 100$ AU. Thus, the case $\Lambda \propto r$ is not realistic. To explain the behaviour of the intensity near solar minimum, it is better to use the theoretical case with $\alpha = -1$ for $r_0/u = 20$ months.

2.7. ANALYSIS FOR SOME STATIONS AND THE TRANSIENT REGION

Dorman and Dorman (1967d) studied the intensity data from four stations: Ottawa, Chicago, Climax, and Huancayo, and obtained theoretical results for $\Lambda = \text{const}$, $\alpha = -\frac{1}{3}$, and $r_0/u = 10$, 20, 30, and 40 months. They noted the following points:

(1) The higher the rigidity R of the particles is, the smaller r_0 is. This can be explained as follows. The maximum size l_{max} of the scattering regions in the sector structure is practically independent of r (since the radial step for one cycle of the Archimedes spiral, 6 AU, is independent of r). Furthermore at great distances, the magnetic field H becomes azimuthal, and the field strength falls off as $H \propto r^{-1}$. Therefore the radius of curvature ϱ of a particle trajectory in a field H, given by the relation $\varrho = R/300\ H$, will increase with r and will eventually equal l_{max}, which has remained constant. The distance at which $\varrho = l_{max}$ will be called r_0^*. Thus, the region $r > r_0^*$ is ineffective for modulation. Using the relations,

$$H(r) \approx 5\left(\frac{r}{r_\oplus}\right)^{-1} \cdot 10^{-5}\ \text{G};\quad r_0^* \approx 225\ l_{max}/R_{\text{ef}}, \tag{13}$$

and, setting for equatorial observations $R_{\text{ef}} = 30$ GV and $l_{max} = 6$ AU, we obtain $r_0^* \approx 255 \times 6/30 \approx 45$ AU. For low energies ($R \lesssim 10$ GV), $r_0^* > r_0$, and the whole region participates in the modulation.

(2) We can explain the additional delay time in the recovery for low-energy cosmic rays by postulating a transition layer at the boundary of the interplanetary medium

with the galactic magnetic field or the region of undisturbed galactic cosmic rays.

This layer is formed in periods of high solar activity as a result of the accumulation and concentration of magnetic inhomogeneities that fail to dissipate into interstellar space. The low-energy particles are scattered by these inhomogeneities; for particles with $R \sim 3$ GV, the mean free path for scattering in the transition region, Λ_{tr}, is estimated to be $\approx 10^{12}$ cm. If the radial thickness of the layer is 100 AU, the corresponding delay time would be ≈ 2 yr. It is also possible that the solar wind magnetic field at $r \sim$ ~ 100 AU, say $H \sim 5 \times 10^{-7}$ G, is compressed and enhanced in the transition layer, so that $H_{tr} \approx 8 \times 10^{-6}$ G, which would exceed the estimated strength of the galactic field, $H_{gal} \approx 3 \times 10^{-6}$ G. The magnetic fields of the inhomogeneities would also be enhanced in the transition layer. It should be noted that in this layer the low-energy particles are accelerated by the Fermi mechanism produced by the chaotic motion of the inhomogeneities.

2.8. Difficulties with Models with a Small Modulation Region

Solar wind models with relatively small modulation regions ($r_0 \lesssim 10 AU$) have been widely discussed in the scientific literature. Some examples are papers by Simpson and Wang (1968), Charakhchyan and Charakhchyan (1968), and Hatton et al. (1968). In these papers monthly averages of the data were used; that is, on the whole they really investigated variations with a relatively short period, which are controlled by relatively small effective modulation regions, $r_0^* \ll r_0$. O'Gallagher and Simpson (1967) and O'Gallagher (1967) used Mariner-4 data to obtain a radial gradient, but Kane and Winckler (1969), after careful elimination of time variation from simultaneous Mariner 4 and OGO data concluded that near solar minimum (cosmic ray maximum) the gradient is $\sim 0\%$ per AU. It must be emphasized that, in the frame of spherically symmetrical models of the solar wind, models with small modulation regions sharply contradict the observations, because they cannot explain the hysteresis phenomenon.

A second difficulty is the following: in order to explain the observed amplitude of the intensity modulation, an average value only one-third as large as that observed for solar cosmic rays is required. If we limit the modulation region to low heliocentric latitude then this second difficulty is aggravated by the introduction of transverse diffusion. There is a third difficulty: in the models with small modulation regions the residual modulation at solar minimum must be very small, which contradicts the results of Pathak and Sarabhai (1970) and the combined analysis of proton and α-particle spectra (see Section 3 below).

3. Spectral Variations in the Low-Energy Range

As was noted above (Section 1) the depth of the modulation in the non-relativistic energy domain depends not only on R but also on v. For particles with different A/Z (e.g. for protons $A/Z = 1$, for α-particles $A/Z = 2$, for electrons $A/Z \approx 5 \times 10^{-4}$, etc.) at the same R (i.e., at the same Λ) v is different, and this fact may be used to determine the modulation parameters (Dorman and Dorman, 1965b):

$$\frac{D_{\text{mod}}(R)}{D_0(R)} = \exp\left[(-a(R)/\beta)\right], \text{ where } a(R) = \frac{3u(r_0 - r_\oplus)}{c\Lambda(R)} \text{ and } \beta = v/c;$$

$$\frac{v_p}{v_\alpha} = \left(\frac{R^2 + 3.52}{R^2 + 0.88}\right)^{1/2}; \quad \frac{D_{\text{mod},p}}{D_{\text{mod},\alpha}} = k \exp\left\{\frac{ca(R)}{v_p}\left[\left(\frac{R^2 + 3.52}{R^2 + 0.88}\right)^{1/2} - 1\right]\right\};$$

$$k = D_{0,p}(R)/D_{0,\alpha}(R) \approx 7. \tag{14}$$

Here, D_{mod} is the modulated flux, as contrasted with D_0, the unmodulated flux in interstellar space. The subscripts p and α refer to protons and α-particles.

We can determine $a(R)$ from experimental data:

$$a(R) = \frac{v_p \ln\left[D_{\text{mod},p}(R)/kD_{\text{mod},\alpha}(R)\right]}{c\left[\left(\frac{R^2 + 3.52}{R^2 + 0.88}\right)^{1/2} - 1\right]}. \tag{15}$$

Now let us consider simultaneous balloon measurements of protons and α-particles made in 1963. From these we can find $(D_{\text{mod},p}(R))/(D_{\text{mod},\alpha}(R))$, calculate $a(R)$, and then obtain $D_0(R)$. In Figure 5, it is apparent that the bend in the spectrum is completely explained by the modulation. Knowing $a(R)$, we can find the predicted gradient and the ratio of the gradients for α-particles and protons (Dorman and Dorman, 1965b):

$$g(r_\oplus) = \left[\frac{dD_{\text{mod}}(R)/dr}{D_{\text{mod}}(R)}\right]_{r=r_\oplus} = \frac{a(R) r_\oplus \cdot 100}{\beta(r_0 - r_\oplus)}, \%/\text{AU} \tag{16}$$

$$g_\alpha(r_\oplus)/g_p(r_\oplus) = \left(\frac{R^2 + 3.52}{R^2 + 0.88}\right)^{1/2}. \tag{17}$$

The predicted gradient is $\sim 10\%$ per AU. The values of the gradient and the gradient ratio agree well with observation. A similar analysis (Dorman and Dorman, 1967e) was carried out for the measurements of protons, α-particles, and electrons in 1964 on the IMP-1 satellite. The results are given in Table I.

This method may be generalized (Dorman and Dorman, 1967d) for measurements of

TABLE I

R, GV	$\dfrac{D_{\text{mod},p}(R)}{D_{\text{mod},\alpha}(R)}$	$a(R)$	$D_{\text{mod}}(R)/D_0(R)$		
			p	α	e
1.0	1.45	0.22	0.75	0.52	0.80
0.9	2.04	0.36	0.60	0.43	0.70
0.8	2.95	0.46	0.49	0.30	0.62
0.7	4.26	0.52	0.42	0.214	0.59
0.6	5.76	0.53	0.37	0.165	0.58
0.5	8.1	0.52	0.33	0.126	0.59

K different kinds of particles (with different A/Z) made in n different time intervals. Thus:

$$D_{\text{mod},\,ij}(R, r) = D_{0i}(R) \exp\left[-a_j(R, r)/\beta_i(R)\right], \quad (1 \leqslant i \leqslant K, 1 \leqslant j \leqslant n),$$
(18)

which is a system of $K \times n$ equations, in which there are $(K+n)$ unknown quantitities, $D_{0i}(R)$ and $a_j(R, r)$. The conditions for the consistency of the system are $(K-1) \times \times (n-1)$ in number, and have the form:

$$\left(D_{\text{mod},\,ij}/D_{\text{mod},\,ij'}\right)^{\beta_i(R)} = \cdots = \left(D_{\text{mod},\,kj}/D_{\text{mod},\,kj'}\right)^{\beta_k(R)}, \quad (j \neq j').$$
(19)

We thus obtain $K+n-1$ independent equations for $K+n$ unknowns. The data from the station network can be used to determine one of the $a_j(R, r)$ and to extrapolate to the low-energy region. This method is very promising, since it permits the spectra in the interstellar medium $D_{0i}(R)$ and the various $a_j(R, r)$ to be determined.

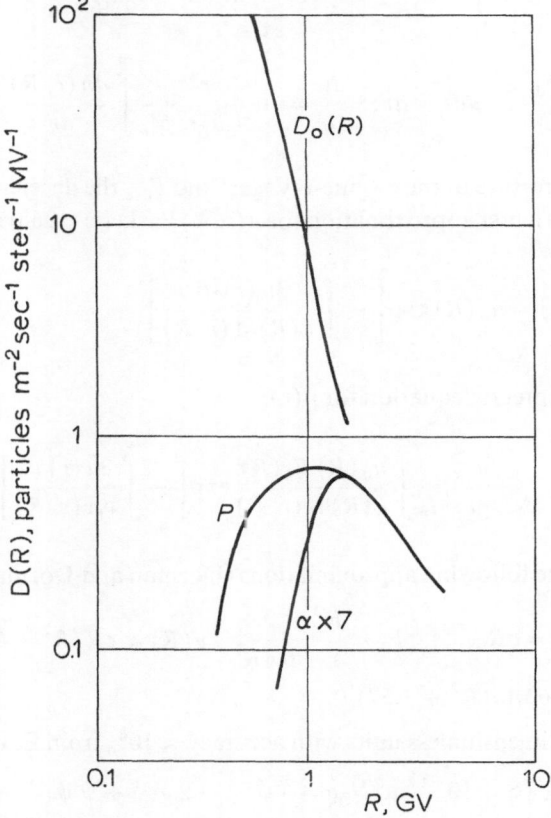

Fig. 5. The observed rigidity spectrum of protons (labelled P) and of α-particles (scaled up by a factor of 7, and labelled $\alpha \times 7$), and the predicted proton spectrum outside the modulation region (labelled $D_0(R)$).

4. The Reciprocal Action of Cosmic Rays on the Solar Wind

If we calculate the total energy density of galactic cosmic rays by using spectra corrected for modulation effects, we conclude (Dorman and Dorman, 1967d) that the total energy density appears to be 2–3 times higher than the usually adopted value (1 eV/cm³ for $E_k > 1$ GeV). Hence, the cosmic-ray pressure is substantial and the reciprocal or reverse effect of cosmic rays on the solar wind should be taken into account. If l and L are the transverse and longitudinal sizes of a volume of plasma, M is its mass, M_p the mass of proton, $N_\pm$ and $u_\pm$ are the concentration and velocity of solar wind near the Earth's orbit, we have, according to Dorman and Dorman (1968b):

$$P = \int_0^\infty n(r, R) E_k \, dR; \quad F_{\text{dec}} = l^2 \int_0^\infty [n(r, R) - n(r + L, R)] E_k \, dR \approx$$

$$- l^2 L \int_0^\infty \frac{dn(r, R)}{dr} E_k \, dR; \quad M = l^2 L N_\pm u_\pm r_\pm^2 u^{-1} r^{-2} M_p; \tag{20}$$

$$\frac{du}{dt} = \frac{F_{\text{dec}}}{M}; \quad u \, dt = dr; \quad \frac{du}{dr} = -\frac{r^2}{N_\pm u_\pm r_\pm^2 M_p} \int_0^\infty \frac{dn(r, R)}{dr} E_k \, dR,$$

in which P is the pressure of the cosmic-ray 'gas' and F_{dec} the decelerating force.

Let us choose as a first approximation for $n(r, R)$ Parker's (1963) solution

$$n(r, R) = n_0(R) \exp\left[-\int_r^\infty \frac{3u(r) \, dr}{v(R) \Lambda(r, R)}\right]. \tag{21}$$

Then we have the integral equation for $u(r)$:

$$\frac{du}{dr} = -\frac{3ur^2}{N_\pm u_\pm r_\pm^3 M_p} \int_0^\infty \frac{n_0(R) E_k(R)}{v(R) \Lambda(r, R)} \exp\left[-\int_r^\infty \frac{3u(r) \, dr}{v\Lambda(r, R)}\right] dR. \tag{22}$$

Let us introduce the following approximations (Dorman and Dorman, 1968c):

$$n_0(R) = aR^{-2.5}; \quad E_k \approx \frac{R^2 \times 0.4}{1 + 0.4R}; \quad v(R) \approx cR(R^2 + 1.57)^{-1/2};$$

$$\Lambda \approx 0.63\Lambda_0(R^2 + 1.57)^{1/2}. \tag{23}$$

Then we have, in dimensionless units with accuracy $\approx 10\%$ from Equation (22):

$$\psi'' = \psi' [5 \times 10^{-12} u_\pm^2 N_\pm \varrho^{-2} \psi' \psi^{-2} + 2\varphi^{-1} + \psi' \psi^{-1} - \lambda_0' \lambda_0^{-1}], \tag{24}$$

where

$$\psi = u_\pm/u, \quad \varrho = r/r_\pm, \quad \lambda_0 = \Lambda_0/r_\pm; \quad \psi(1) = 1,$$

$$\psi'(1) = 1.13 \times 10^{-6} (u_\pm/3 \times 10^7)^{-5} (N_\pm/5)^{-2} \lambda_0$$

A solution for ψ was found by Dorman and Dorman (1968d) for three regions:

(1) $\varrho < \varrho_1$, $\psi = 1 - \dfrac{\psi'(1)}{3}(\varrho^3 - 1)$;

(2) $\varrho_1 < \varrho < \varrho_2$, a numerical solution for ψ; (25)

(3) $\varrho > \varrho_2$, $\psi = \psi(\varrho_2)\exp\left[-\dfrac{\varrho - \varrho_2}{\varrho_0}\right]$.

If the cosmic-ray energy density $W_{CR} \approx 1$ eV cm^{-3}, then $u(r)$ decreases 10% at $r \approx 100$ AU, 50% at $r \approx 200$ AU, and 90% at $r \approx 400$ AU. If $W_{CR} \approx 3$–4 eV cm^{-3}, then $u(r)$ decreases 10% at $r \approx 50$ AU, 50% at $r \approx 100$ AU, and 90% at $r \approx 200$ AU. If we assume that $n_0(E_k) = aE_k^{-2.5}$, then $W_{CR} = \int_{E_{k\,min}}^{\infty} n_0(E_k)E_k dE_k$; in this case the solar wind velocity decreases by a factor of 10 in the distance r_0 (i.e. $u(r_0) = 0.1\,u_{\ast}$). Some results are summarized in Table II (Dorman and Dorman, 1969). From Table II it can be seen that the most realistic values of $E_{k\,min}$ are 0.1 and 0.03 GeV, which gives reasonable values for the delay time, T_{delay}.

TABLE II

$E_{k\,min}$, GeV	W_{CR}, eV cm^{-3}	r_0, AU	T_{delay}, Years
0.5	1.4	240	3.9
0.1	3.2	116	1.9
0.03	5.8	74	1.2

If in addition to the radial gradient of cosmic rays, transverse gradients exist, transverse forces from cosmic rays on the magnetized plasma will appear. For example, imagine a solar wind jet with an enhanced flux of magnetic inhomogeneities (i.e. small Λ). Inside this jet the modulation will be greater than outside, the cosmic-ray density will be lower, and a transverse compression from cosmic-ray pressure (perpendicular to the Sun's equatorial plane) will appear. This will result in a further decrease in Λ, and hence in further enhancement of the modulation and an even greater transverse compression, i.e. the jets will be focussed (see Figure 6).

We have

$$\frac{du_\perp(r)}{dr} \approx \frac{r[W_{CR}(r_0) - W_{CR}(r)]}{\theta_0 M_p N_{\ast} r_{\ast}^2 u_{\ast}},$$ (26)

where $u_\perp$ is the transverse velocity of the plasma and θ_0 the effective heliocentric latitude. From Equation (26) it is possible to find the relative transverse shift of the plasma, $\Delta/r\theta_0$.

Table III summarizes the results of the calculating $\Delta/r\theta_0$ for $\theta_0 = 30°$ and for various values of $E_{k\,min}$, r, and $u_\perp$. Each box gives r (in AU) on the left, $u_\perp$ (in cm sec^{-1}) on the upper right, and $\Delta/r\theta_0$ on the lower right. The Table shows that at r greater than several AU the effect of the transverse pressure becomes appreciable, even for $E_{k\,min} = 0.03$ GeV.

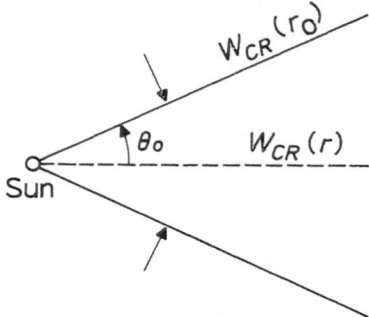

Fig. 6. Schematic representation of the focussing of solar wind jets when transverse gradients of cosmic-ray intensities are present,

TABLE III

$E_{k\,min}$, GeV	Solar maximum					
0.5	7.2	$1.9 \cdot 10^5$	24	$1.6 \cdot 10^6$	72	10^7
		$3.5 \cdot 10^{-3}$		$2.8 \cdot 10^{-2}$		0.2
0.03	2.2	$1.3 \cdot 10^5$	7.2	$1.3 \cdot 10^6$	22.2	$9 \cdot 10^6$
		$3.5 \cdot 10^{-3}$		0.02		0.15
	Solar minimum					
0.5	6.9	$1.7 \cdot 10^5$	23	$1.4 \cdot 10^6$	69	$8 \cdot 10^6$
		$4.2 \cdot 10^{-3}$		$3.4 \cdot 10^{-2}$		0.27
0.03	1.4	$1.0 \cdot 10^5$	4.6	$1 \cdot 10^6$	13.8	$6.6 \cdot 10^6$
		$2.4 \cdot 10^{-3}$		$2.4 \cdot 10^{-2}$		0.18

We conclude that, when studying cosmic-ray modulation at great distances from the Sun, it is necessary to take account of the reverse effect of cosmic rays on the solar wind, i.e., to solve the non-linear problem of galactic cosmic ray modulation.

5. Modulation Models without Spherical Symmetry

If we suppose that the modulation region is small ($r_0 \lesssim 10$ AU), then the hysteresis phenomenon in the 11-yr cosmic-ray variation can be explained only if we abandon a spherically symmetrical geometry for the solar wind. One kind of non-spherical model attempts to explain the hysteresis by the shift of the effective heliocentric latitude for solar activity from high to low latitudes during the 11-yr cycle (Quenby, 1965; Kudo and Wada, 1968; Stozhkov and Charakchyan, 1969). But then one would expect to find a large transverse (north-south) cosmic ray gradient in interplanetary space – in fact, an order of magnitude greater than that derived from seasonal cosmic-ray variations by Dorman *et al.* (1967a, b). Because of the sharp north-south asymmetry of solar activity, one would not expect to find that the minimum value of this gradient will always coincide with the neighborhood of the Earth's orbit. A very small value of the transverse gradient also resulted from an experimental investigation by Dorman and Inosemtseva (1970) of three-dimensional cosmic-ray anisotropy and its interpretation on the basis of on anisotropic diffusion model for cosmic rays.

A more detailed and careful analysis of the effect of the shift of the effective heliocentric latitude for solar activity was carried out by Gushchina *et al.* (1970a, b, c). They introduced the so-called *HL*-index of solar activity, which takes into account the latitude distribution of solar activity and the variation of the Earth's heliocentric latitude with time:

$$\overline{HL} = q \int_{-\pi/2}^{\pi/2} A(\theta) \exp\left[-\left|\frac{\theta - \theta_{\oplus}}{\theta_0}\right|\right] d\theta, \tag{27}$$

In which q is a normalizing factor, $A(\theta)$ is an arbitrary index of solar activity varying with the heliocentric latitude θ, $\theta_{\oplus}$ is the Earth's latitude, θ_0 is the characteristic semiangle of the active zone, in which occur the processes that affect the interplanetary medium. Gushchina *et al.* (1970) showed that for $\theta_0 = 5$–$10°$ the hysteresis vanishes; and according to Pimenov *et al.* (1970) at $\theta_0 = 5$–$10°$ the theoretical amplitude of the cosmic-ray variations is more than 10 times greater than the observed variation. Agreement between theoretical and observational values of the seasonal amplitudes is found only at $\theta_0 \approx 30$–$50°$; in this case, however, the hysteresis is approximately the same as that in which a solar activity index for the whole solar disc is used. But, as we have already seen in Section 2, in this case the modulation region must be large, which contradicts the smaller model assumed in the present discussion.

Pathak and Sarabhai (1970) carried out an interesting investigation of a model in which the solar wind is concentrated to low latitudes. Using data on the intensity of the green coronal line $\lambda = 5303$ Å at latitudes $\pm 5°$, they employed the following relation to estimate the temperature of the inner solar corona T_0 in the region $\pm 5°$.

$$T_0 = 0.18 \times 10^6 \, I \, (5303 \text{ Å}) + 1.03 \times 10^6. \tag{28}$$

This relation was obtained by Billings and Hatt (1968) by determining T_0 for the whole solar disk from the ratio of the intensities of the green, red ($\lambda = 6374$ Å), and yellow ($\lambda = 5694$ Å) coronal emission lines. Using the relation between T_0 and solar wind velocity u for the hydrodynamic expansion model derived by Parker (1963),

$$u = \left[\frac{4kT_0}{M_p}\left(\frac{\alpha}{\alpha - 1}\right)\right]^{1/2}, \tag{29}$$

where k is the Boltzmann constant, M_p is the mass of the protons, α is the polytropic index ($\alpha > 1.0$), and T_0 is the temperature at the base of the corona, Pathak and Sarabhai (1970) determined u for the region $\pm 5°$ for the period 1957–67. If the size of the modulation region is r_0, and if the mean free transfer path for particle scattering

Λ and the solar wind velocity u do not depend on r, then we can write

$$\ln n\,(r_{\hbox{\scriptsize\male}}, Ze, R, t) = \ln n_0\,(Ze, R) - \frac{3\,(r_0 - r_{\hbox{\scriptsize\male}})}{v\Lambda}\cdot u\,(t) \tag{30}$$

which follows from Equation (1).

Figures 6 and 7 illustrate the relation between u and the cosmic-ray intensity at Deep River ($R_{\text{ef}} \approx 10$ GV), Mount Norikura ($R_{\text{ef}} \approx 25$ GV) and Thule (balloon measurements, $R_{\text{ef}} \approx 3$ GV). These figures show that the relation between $\ln n$ and $u(t)$ is

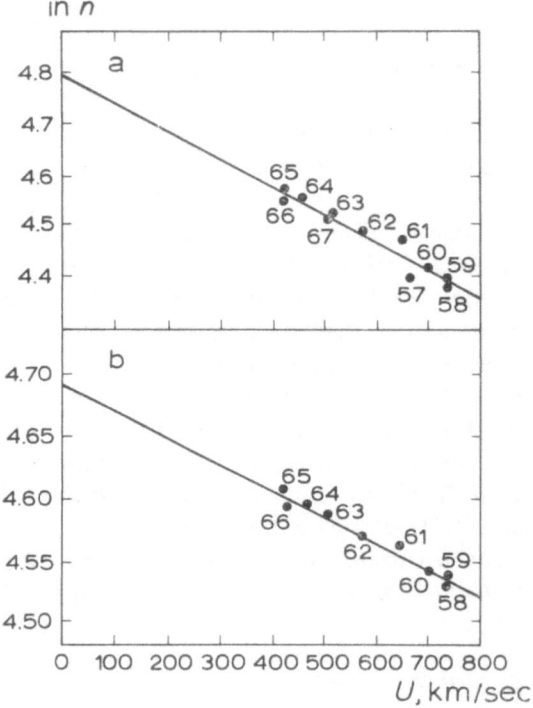

Fig. 7. Correlations of observed cosmic-ray intensities $\ln n$ with solar wind velocities u, constituting an observational verification of Equation (30): (a) yearly mean neutron intensities at Deep River, for the period 1957–1967; (b) yearly mean neutron intensities at Mount Norikura for the period 1958–1966. The yearly mean solar wind velocities have been derived with $\alpha = 1.10$ in Equation (29).

linear, i.e. that the coefficient of $u(t)$, $(3\,(r_0 - r_{\hbox{\scriptsize\male}})/v\Lambda)$, is constant. On the other hand, it is well known (for example from solar cosmic ray data of Dorman and Miroshnichenko, 1968) that Λ decreases considerably from solar minimum to maximum – up to possibly one order of magnitude. This means that the quantity $r_0 - r_{\hbox{\scriptsize\male}}$ must also decrease by about the same amount, that is r_0 must decrease with an increase of solar activity. But this conclusion contradicts the supposition that the size of the modulation region is determined by the interaction of the solar wind with the galactic magnetic and cosmic-ray field. Pathak and Sarabhai (1970) argue that the increase of solar activity leads to solar wind turbulization at a smaller distance, that is, that the size of the region with a *regular* magnetic field decreases. But let us consider another possibility. Equation (30) is valid only for the spherically symmetrical cases, but Pathak and Sarabhai (1970) are considering a model in which the solar wind is concentrated to

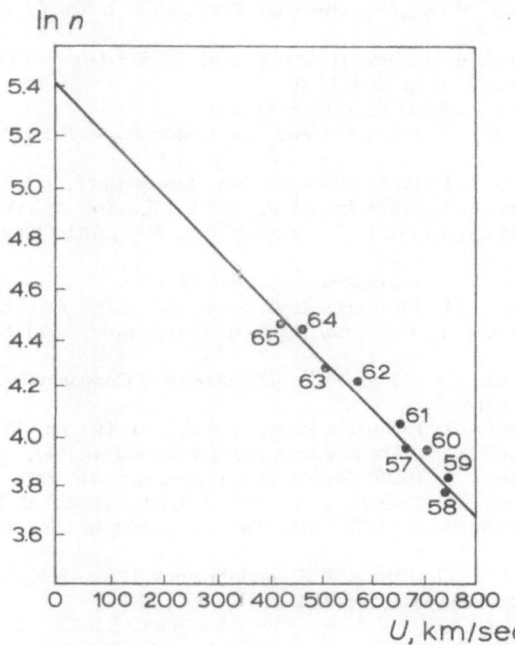

Fig. 8.   Correlation of yearly mean cosmic-ray intensities $\ln n$, obtained by Neher and Anderson (1966) for the years 1957–1965 with ion-chamber measurements at the top of the atmosphere over Thule, with the yearly mean solar wind velocity u. (See Figure 7 for other details.)

low latitudes. In order to take this fact into account, let us introduce into the coefficient of $u(t)$ an additional factor A_1, which reflects transverse diffusion. The value A_1 can vary strongly with solar activity, because of the shift of the effective latitude of solar activity. If A_1 strongly decreases with increasing solar activity (because the transverse diffusion increases), then, even with a corresponding increase in r_0, and hence in $3(r_0 - r_{\oplus})$, and a decrease in Λ, the entire coefficient of $u(t)$ *can* remain constant, in agreement with the experimental result of Pathak and Sarabhai (1970).

References

Axford, W. I.: 1965, *Planetary Space Sci.* 13, 115.
Axford, W. I. and Newman, R. C.: 1965, *Proc. Int Conf. Cosmic Rays* 1, 173.
Belov, A.V. and Dorman, L. I.: 1969, *Geomagnetizm i Aeronomiya* 9, 972.
Belov, A. V. and Dorman, L. I.: 1971, *Geomagnetizm i Aeronomiya*, 11, 328.
Billings, D. E. and Hatt, W. A.: 1968, *Astrophys. J.* 151, 743.
Blokh, G. M., Blokh, Ya. L., and Dorman, L. I.: 1964, *Izv. Nauk, S.S.S.R.* (*ser. fiz.*) 28, 1985.

Blokh, Ya. L., Dorman, L. I., and Kaminer, N. S.: 1960, *Proc. Int. Conf. Cosmic Rays*, Akad. Nauk S.S.S.R. Moscow **4**, 178.

Blokh, Ya. L., Dorman, L. I., and Kaminer, N. S.: 1961, *Col. Papers Cosmic Rays*, Akad. Nauk S.S.S.R. Moscow, **4**, 59.

Blokh, Ya. L., Glokova, E. S., and Dorman, L. I.: 1959, *Col. Papers Cosmic Rays*, Akad. Nauk S.S.S.R. Moscow **1**, 7.

Charakhchyan, A. N. and Charakhchyan, T. N.: 1968, *Geomagnetizm i Aeronomiya* **8**, 29.

Coleman, P. J.: 1967, Inst. of Geophys. and Plan. Phys., Univ. of California (Los Angeles) Publ. No. 629 (preprint).

Dolginov, A. S. and Toptigin, I. N.: 1968, *Izv. Akad. Nauk S.S.S.R. (ser. fiz)* **32**, 182.

Dorman, I. V.: 1970, Thesis, Moscow State Univ.

Dorman, I. V.: 1971, *Geomagnetizm i Aeronomiya* **11**, 330.

Dorman, I. V. and Dorman, L. I.: 1965a, *Col. Papers on Cosmic Rays*, Akad. Nauk S.S.S.R., Moscow **7**, 5.

Dorman, I. V. and Dorman, L. I.: 1965b, *Geomagnetizm i Aeronomiya* **5**, 666.

Dorman, I. V. and Dorman, L. I.: 1967a, *Izv. Akad. Nauk S.S.S.R. (ser. fiz.)* **31**, 1239.

Dorman, I. V. and Dorman, L. I.: 1967b, *Col. Papers Cosmic Rays*, Akad. Nauk S.S.S.R., Moscow **8**, 65.

Dorman, I. V. and Dorman, L. I.: 1967c, *J. Geophys. Res.* **72**, 1513.

Dorman, I. V. and Dorman, L. I.: 1967d, *Izv. Akad. Nauk S.S.S.R. (ser. fiz.)* **31**, 1273.

Dorman, I. V. and Dorman, L. I.: 1967e, *Col. Papers on Cosmic Rays*, Akad. Nauk S.S.S.R. Moscow **8**, 100.

Dorman, I. V. and Dorman, L. I.: 1968a, *Proc. of 5th School of Cosmophysics, Apatiti* (Kolskiy filial Akad. Nauk S.S.S.R.), p. 183.

Dorman, I. V. and Dorman, L. I.: 1968b, *Izv. Akad. Nauk S.S.S.R. (ser. fiz.)* **32**, 1838.

Dorman, I. V. and Dorman, L. I.: 1968c, *Geomagnetizm i Aeronomiya* **8**, 817.

Dorman, I. V. and Dorman, L. I.: 1968d, *Geomagnetizm i Aeronomiya* **8**, 1000.

Dorman, I. V. and Dorman, L. I.: 1969, *Izv. Akad. Nauk S.S.S.R. (ser. fiz.)* **33**, 1908.

Dorman, I. V. and Dorman, L. I.: 1970, *Acta Physica Academiae Scientiarum Hungaricae* **29**, Suppl. 2, 17.

Dorman, I. V., Dorman, L. I., Gushchina, R. T., and Pimenov, I. A.: 1969, *Proc. All-Union Conf. on Cosmic Ray Physics* (Tashkent, 1968) part II, **2**, 29.

Dorman, L. I.: 1960, *Proc. Int. Conf. on Cosmic Rays*, Akad. Nauk S.S.S.R., Moscow, **4**, 328.

Dorman, L. I.: 1961, *Col. Papers on Cosmic Rays*, Akad. Nauk S.S.S.R. Moscow, **4**, 251.

Dorman, L. I.: 1963a, in *Progress in Elementary Particle and Cosmic Rays Physics* (ed. by J. G. Wilson and S. A. Wouthuysen), North-Holland Publ. Co., Amsterdam, **7**, 1.

Dorman, L. I.: 1963b, *Cosmic Ray Variation and Space Exploration, Akad. Nauk S.S.S.R.*, Moscow.

Dorman, L. I.: 1963c, *Proc. Int. Conf. on Cosmic Rays, Jaipur (India)* **1**, 80.

Dorman, L. I.: 1965, *Proc. Int. Conf. on Cosmic Rays, London* **1**, 229.

Dorman, L. I.: 1966, *Izv. Akad. Nauk S.S.S.R. (ser. fiz.)* **30**, 1722.

Dorman, L. I.: 1967, *Col. Papers on Cosmic Rays*, Akad. Nauk S.S.S.R., Moscow **8**, 305.

Dorman, L. I. and Inosemtseva, O. I.: 1970, *Izv. Akad. Nauk S.S.S.R. (ser. fiz.)* **34**, 2386.

Dorman, L. I. and Kaminer, N. S.: 1970, *Ann. Géophys.* **26**, N3, 697.

Dorman, L. I. and Kobylinski, Z.: 1968, *Izv. Akad. Nauk S.S.S.R. (ser. fiz.)* **32**, 1928.

Dorman, L. I. and Miroshnichenko, L. I.: 1968, *Solar Cosmic Rays*, Akad. Nauk S.S.S.R., Moscow.

Dorman, L. I., Kaminer, N. S., and Kebuladze, T. V.: 1969a, *Geomagnetizm i Aeronomiya* **9**, 217.

Dorman, L. I., Kaminer, N. S., and Kebuladze, T. V.: 1969b, *Geomagnetizm i Aeronomiya* **9**, 617.

Dorman, L. I., Kaminer, N. S., and Kebuladze, T. V.: 1970, *Izv. Akad Nauk S.S.S.R. (ser. fiz.)* **34**, 2360.

Dorman, L. I., Lusov, A. A., and Mamrukova, V. P.: 1967a, *Doklady Akad. Nauk S.S.S.R.* **172**, 839.

Dorman, L. I., Lusov, A. A., and Mamrukova, V. P.: 1967b, *Izv. Akad. Nauk S.S.S.R. (ser. fiz.)* **31**, 1368.

Gleeson, L. J.: 1968, *Proc. Astron. Soc. Australia* **1**, 130.

Gleeson, L. J. and Axford, W. I.: 1967, *Astrophys. J. Letters* **149**, 115.

Gleeson, L. J. and Axford, W. I.: 1968, *Canad. J. Phys.* **46**, part 4, 937.

Gushchina, R. T., Dorman, I. V., Dorman, L. I., and Pimenov, I. A.: 1968, *Izv. Akad. Nauk S.S.S.R. (ser. fiz.)* **32**, 1924.

Gushchina, R. T., Dorman, I. V., Dorman, L. I., and Pimenov, I. A.: 1969, *Proc. All-Union Conf. on Cosmic Ray Physics (Tashkent, 1968)*, part II, **2**, 29.

Gushchina, R. T., Dorman, I. V., Dorman, L. I., and Pimenov, I. A.: 1970a, *Acta Physica Academiae Scientiarum Hungaricae* **29**, Suppl. 2, 219.

Gushchina, R. T., Dorman, I. V., Dorman, L. I., and Pimenov, I. A.: 1970b, *Izv. Akad. Nauk S.S.S.R. (ser. fiz.)* **34**, 2426.

Gushchina, R. T., Dorman, L. I., Ilgach, C. F., Kaminer, N. S., and Pimevon, I. A.: 1970c, *Izv. Akad. Nauk S.S.S.R. (ser. fiz.)* **34**, 2434.

Hatton, C. J., Marsden, P. L., and Willets, A. C.: 1968, *Canad. J. Phys.* **46**, 915.

Heppner, J. P., Ness, N. F., Shilman, T. L., and Scearce, C. S.: 1962, *J. Phys. Soc. Japan*, Suppl. A-II, **17**, 546.

Jokipii, J. R.: 1966, *Astrophys. J.* **146**, 480.

Jokipii, J. R. and Coleman, P. J.: 1968, *J. Geophys. Res.* **73**, 5495.

Kane, S. R. and Winckler, J. R.: 1969, *J. Geophys. Res.* **74**, 6247.

Kodama, M.: 1968, *Japanese Antarctic Research Exped.*, Tokyo, Scient. Rep. No. 5 (Ser. A), August, 1968.

Krymsky, G. F.: 1964, *Geomagnetizm i Aeronomiya* **4**, 977.

Krymsky, G. F.: 1966, Dissertation, Moscow State Univ., Moscow.

Krymsky, G. F.: 1969, *Modulation of Cosmic Rays in Interplanetary Space*, Akad. Nauk S.S.S.R., Moscow.

Kudo, S. and Wada, M.: 1968, *Rep. Ionosphere Space Res. Japan* **22**, 137.

Neher, H. V. and Anderson, H. R.: 1966, *Proc. Int. Conf. Cosmic Rays, London* **1**, 153.

Ness, N. F.: 1966, *J. Geophys. Res.* **71**, 3319.

O'Gallagher, J. J.: 1967, *Astrophys. J.* **150**, 675.

O'Gallagher, J. J. and Simpson, J. A.: 1967, *Astrophys. J.* **147**, 819.

Parker, E. N.: 1958, *Phys. Rev.* **110**, 1445.

Parker, E. N.: 1963, *Interplanetary Dynamical Processes*, John Wiley, New York.

Parker, E. N.: 1965a, *Proc. Int. Conf. on Cosmic Rays, London* **1**, 26.

Parker, E. N.: 1965b, *Planetary Space Sci.* **13**, 9.

Pathak, P. N. and Sarabhai, V. A.: 1970, *Planetary Space Sci.* **18**, 81.

Quenby, J. J.: 1965, *Proc. Int. Conf. Cosmic Rays, London* **1**, 3.

Roelof, E. C.: 1966, PhD. Thesis, Univ. of California, Berkeley.

Simpson, J. A.: 1963, *Proc. Int. Conf. Cosmic Rays, Jaipur* **2**, 155.

Simpson, J. A. and Wang, W. R.: 1967, *Astrophys. J.* **149**, 73.

Singer, S. F.: 1958, *Nuovo Cimento* **8**, Suppl. 2, 334.

Sousk, S. F. and Lenchek, A. M.: 1969, *Astrophys. J.* **156**, 1107.

Stozhkov, Yu. I. and Charakhchyan, T. N.: 1969, *Geomagnetizm i Aeronomiya* **9**, 803.

Tverskoy, B. A.: 1968, *Zh. Eksp. i Teor. Fizik* **53**, 1417.

LOW-ENERGY COSMIC RAYS IN INTERPLANETARY SPACE

S. N. VERNOV and G. P. LYUBIMOV

Institute of Nuclear Physics, Moscow State University, Moscow, U.S.S.R.

Abstract. The report presents the data of measurements of low-energy cosmic rays (mainly protons with energies $E_p \approx 1$–5 MeV and $E_p > 30$ MeV) from Zond-3 and Venus-2, -3, -4, -5, and -6 space probes for the period 1965–69. During this period the frequency of occurrence, duration, and intensity of the solar cosmic ray flares increased. Various spatial forms of the solar cosmic ray increase were found and studied. It has been shown from observations of solar cosmic rays injected by solar flares in 1965–67 that the low-energy protons propagate from the Sun in two ways: (1) rapidly, by means of particle movement along the magnetic force lines, and (2) slowly, by means of particle transfer in the layer corresponding to the falling phase of the Forbush-decrease, i.e. between the shock wave-front and the front of the hot plasma ejected by a flare. Reflection of particles from some 'mirror' region and considerable deceleration of shock waves were found in the first and second cases, respectively. It has been shown from the results of comparison of the galactic cosmic-ray decreases with active regions on the Sun that the amplitude of the decrease is determined by the dynamics of active regions, by their latitude, and their phase of development. Characteristic features of the results of 1969 include: the presence of very large and enduring proton fluxes with $E_p > 30$ MeV observed following the shock-wave fronts, a harder spectrum, and complex increases caused by groups of powerful solar flares. Considerable depth and rapid intensity-recovery are features of Forbush decreases in this period while the study of mean velocities of shock wave propagation indicates a pronounced direction of flare ejections and sustained pushing of shock waves by flare gases.

1. Introduction

The study of cosmic rays in interplanetary space at various distances from the Sun and during various periods of solar activity presents a possibility of observing the connection between local events (in time and space) and enduring cyclic changes on the Sun, in interplanetary space, and in cosmic rays. Various phases of the solar activity cycle differ strongly from each other. The difference is also strong within individual forms of solar activity displays: slow processes in active regions, and rapid processes in flares. The difference is also strong between different active regions and between different flares. At the same time, however, there is much in common between these displays of activity.

In accordance with the variety of the events on the Sun the characteristics of the cosmic ray fluxes also vary. However, when studying carefully the whole variety of events one can find much common in them.

The property of cosmic rays of propagating to great distances and undergoing the influence of magnetic fields permits the large-scale characteristics of interplanetary space to be studied within a large volume, and the study of cosmic rays of various energies permits one to vary the resolution of this method. Success of the study is determined by correct finding of the correspondence between events on the Sun, in interplanetary space, and in cosmic rays.

Examined below in the report will be the characteristics of the low-energy cosmic-ray fluxes of solar and galactic origin depending on solar activity, mainly in 1969. The data obtained from Zond-3 and Venus-2, -3, -4, -5, and -6 during 1965–69 will be used.

Dyer (ed.), Solar Terrestrial Physics/1970: Part II, 92–109. All Rights Reserved.

2. General Characteristics of Cosmic Ray Fluxes During 1965–69

Earlier works by Vernov *et al.* (1967, 1968a) reported the data on the 11-yr cycle of intensity variations of galactic cosmic rays with energies $E_p > 30$ MeV from 1958–1965. It was shown that the galactic cosmic-ray flux had increased during the descending branch of the solar cycle. According to the data from Zond-3 and Venus-2, the radial gradient of galactic cosmic rays was $+5\%$ per 1 AU in 1965. The measurements during the flights of Venus-4 in 1967 (Vernov *et al.*, 1969a) and Venus-5 and -6 in 1969 (Vernov *et al.*, 1969b) during the ascending branch of the solar cycle showed a decrease in galactic cosmic-ray flux after its peak, which occurred in 1965–66. The cosmic-ray radial gradient, as determined from data of Venus-4 and Explorer-34 in 1967, changed its sign and was -13.5% per 1 AU (Vernov *et al.*, 1969b).

Further changes in time, from 1965–1969, in fluxes of both galactic and solar cosmic rays will be described below.

Figure 1 presents, as a function of time, the daily averages of the intensity of: (1) protons with energy $1 < E_p < 5$ MeV, (2) electrons with $E_e > 0.05$ MeV, (3) protons with $E_p > 30$ MeV measured on space probes, as well as those of intensity of the galactic cosmic-ray neutron component according to ground measurements at Deep River station (ESSA, 1965–69). These data cover the period from July 1965 to May 1969. The figure illustrates the time dependence of solar and galactic cosmic rays, i.e. the dependence on the solar activity level. Consider three successive intervals: the first (Zond-3 and Venus-2) July 1965 – January 1966; the second (Venus-4) June 1967 – October 1967; the third (Venus-5 and -6) January 1969 – May 1969.

The proton flux of solar origin with $1 < E_p < 5$ MeV was very strongly increased from the first to the third period. An increase in the number of individual events, their amplitudes, and durations was observed. In the second half on the third period, individual fluxes overlapped. Sometimes the solar cosmic rays of low energies which failed to leave the modulation region were accumulated in this region and could form a density gradient directed to the Sun.

The electron fluxes with $E_e > 0.05$ MeV displayed the same regularities. It should be noted only that no unambiguous connection was found between the values of the proton fluxes and electron fluxes when observed simultaneously.

During the first and second periods, the solar proton fluxes with $E_p > 30$ MeV were present only in the form of individual short bursts (October, 1965; January, 1966; 5 July, 1 August, 25 September, 10 October, 1967; all < 1 day) with an amplitude of several percent; during the third period the intensities and durations of these fluxes increased very abruptly. Besides that, in two cases (on 25 February and 30 March, 1969) bursts of protons with $E_p > 1$ GeV were observed (with neutron monitors). An additional analysis of the > 30 MeV proton level in 1967 (Vernov *et al.*, 1970d) and the reverse sign of the cosmic-ray radial gradient during this period (Vernov *et al.*, 1969b) confirm the increase in the proton flux of these energies.

From 1965–1969 the rate of increase in the proton flux with $E_p > 30$ MeV is higher than that for the protons with energies $1 < E_p < 5$ MeV (see Figure 3); i.e. the solar

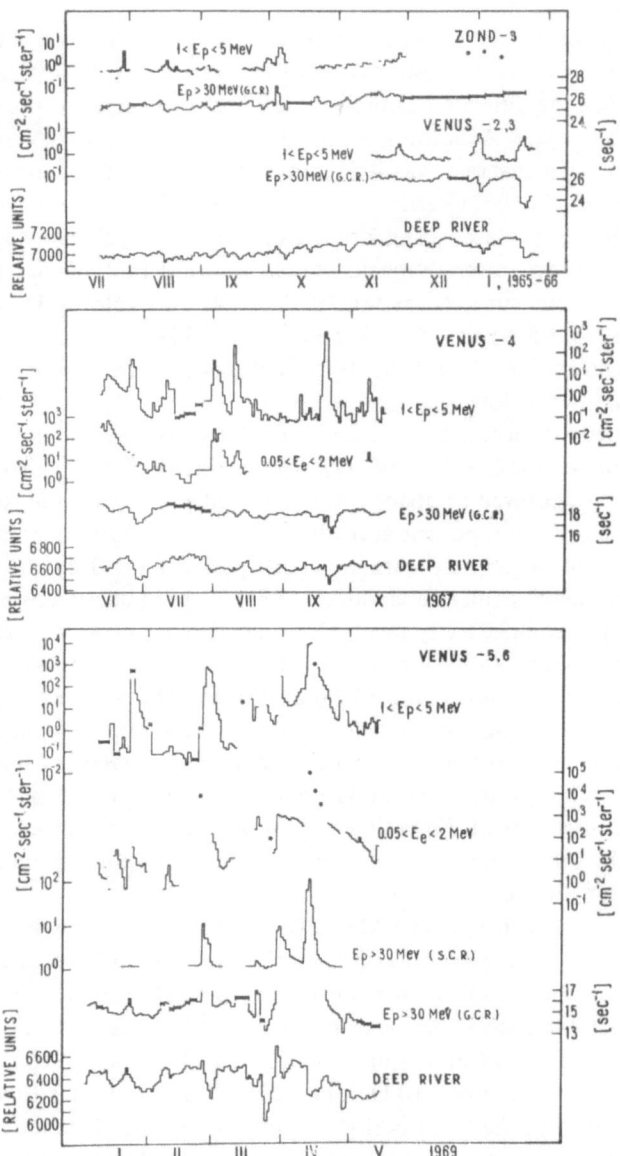

Fig. 1. Mean daily values of the cosmic-ray fluxes at various times are shown, plotted against time scales. The first period corresponds to the measurements on Zond-3 and Venus-2 and -3; the second to those on Venus-4, and the last to Venus-5 and -6. Shown are the fluxes of: (1) protons with $1 < E_p < 5$ MeV; (2) electrons with $0.05 < E_e < 2$ MeV; (3) protons with $E_p > 30$ MeV of galactic and solar origin (labelled GCR and SCR respectively), as well as the neutron monitor data from Deep River.

cosmic-ray spectrum in the 1–30 MeV energy range became harder with time. The same result was obtained for individual solar cosmic-ray bursts (Vernov *et al.*, 1969b, 1970a; Lyubimov, 1969). This effect was probably connected with features of the solar cosmic-ray generation mechanism during that solar activity period.

The level of galactic cosmic rays (i.e. protons with energies $E_p > 30$ MeV and neutron component of cosmic rays in the atmosphere, except for the above-mentioned burst increases) decreased from the first to the third period. In this case the decrease for soft galactic cosmic rays was more rapid than that for hard ones, i.e. the galactic cosmic-ray spectrum became harder. It will be noted that spatial variations of the galactic cosmic-ray flux due to the radial gradient of the galactic cosmic-ray intensity are superposed on these time variations (since the space probes either approach or move away from the Sun during their flight); the space variations, however, may be largely ignored because the time dependence is considerably stronger.

An increase in the number and depth of Forbush decreases from the first to third periods may be pointed out. In this case it should be noted that the increase of the depth of Forbush decreases with time was not great because the galactic cosmic-ray spectrum became simultaneously harder.

This dependence is shown more clearly in Figure 2. This figure presents, as a function of time, the characteristics of Forbush decreases caused by strong solar flares. These events were detected on Earth and in interplanetary space during the flights of (1) Zond-3 and Venus-2 and -3 in 1965–66, (2) Venus-4 in 1967 and (3) Venus-5 and -6 in 1969. Forbush decreases in interplanetary space were partially 'filled' by localized solar cosmic rays; the plot (crosses) presents the data on the depth of Forbush decreases from the neutron monitor at Deep River (ESSA, 1965–69). The mean velocities of shock waves (black points) which caused these Forbush decreases have been calculated from the delay of sudden commencements (ESSA, 1965–69) with respect to the corresponding solar flares (Vernov *et al.*, 1969b, 1969c, 1970a; Lyubimov, 1968, 1969). The ground-based data have also been used to eliminate the dependence of the mean velocity of shock waves and Forbush decrease depth on distance (Lyubimov, 1968, 1969). The same figure shows (open circles) the maximum fluxes of protons with $1 < E_p < 5$ MeV located between the fronts of shock waves and plasma as measured in interplanetary space (Vernov *et al.*, 1969b, 1969c, 1970a). For comparison, the thin stepped line in Figure 2 shows the sunspot number R_z (ESSA, 1965–69). It can be seen that, although the presented parameters of burst disturbances have a large scatter of individual characteristics within each group, on the average they are time-dependent on solar activity; i.e. (1) the depth of Forbush decreases in galactic rays, (2) the mean velocity of shock waves, and (3) the fluxes of localized protons all increased on the average from 1965–1969.

Figure 3 also illustrates the time dependence, i.e. the dependence on solar activity, of the mean values of cosmic-ray fluxes during the flights of (1) Zond-3 and Venus-2 and -3, (2) Venus-4, and (3) Venus-5 and -6. Shown as horizontal bars are the mean fluxes of (1) protons with $1 < E_p < 5$ MeV (thick line); (2) electrons with $E_e > 0.05$ MeV (thin line); (3) solar protons with $E_p > 30$ MeV (thick broken line); (4) galactic protons with $E_p > 30$ MeV (thin broken line); (5) galactic cosmic rays with energy > 1 GeV measured with the neutron monitor at Deep River (dotted line) and expressed in relative units with respect to the level of 1965–66.

Figure 3 confirms the conclusion drawn above about a strong dependence of the solar cosmic-ray fluxes on solar activity. From 1965–1969 the proton fluxes with energy $1 < E_p < 5$ MeV and electron fluxes with $E_e > 0.05$ MeV increased by a factor of about 110 and 580, respectively. The increase of these fluxes may be approximated by exponential functions with time constants of 0.7 and 0.5 yr, respectively.

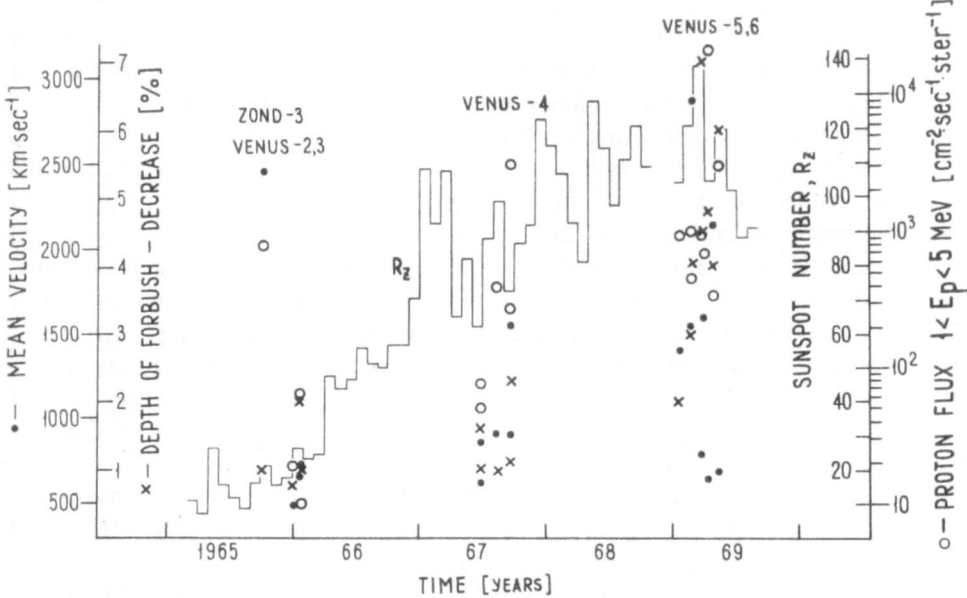

Fig. 2.   Time dependence of Forbush decrease characteristics for three periods, corresponding to the flights of (1) Zond-3, Venus-2 and -3; (2) Venus-4; (3) Venus-5 and -6. Crosses indicate the depths of the galactic cosmic-ray Forbush decreases as shown in the data from Deep River (inner scale, left). The points show the distribution of shock wave velocities, calculated from the delay of sudden commencements (SC) with respect to the peaks of the solar flares corresponding to them (outer scale, left). The open circles are the fluxes of protons with $1 < E_p <$ MeV localised behind the shock wave fronts according to the space probe data (outer scale, right). The stepped line is the mean monthly sunspot number, R_z (inner scale, right).

During the same period the solar proton flux with $E_p > 30$ MeV increased by a factor of about 1700, the increase from 1967–1969 being decidedly abrupt.

During the same period the level of galactic protons with $E_p > 30$ MeV decreased by a factor of 1.7, the decrease from 1967–1969 becoming slow.

It will be noted at the end that according to the data from Venus-6 and Explorer-34 (ESSA, 1965–69) in January-May 1969 the radial gradient for protons with $E_p > 30$ MeV determined in the same way as in findings by Vernov et al. (1969b) was on the average

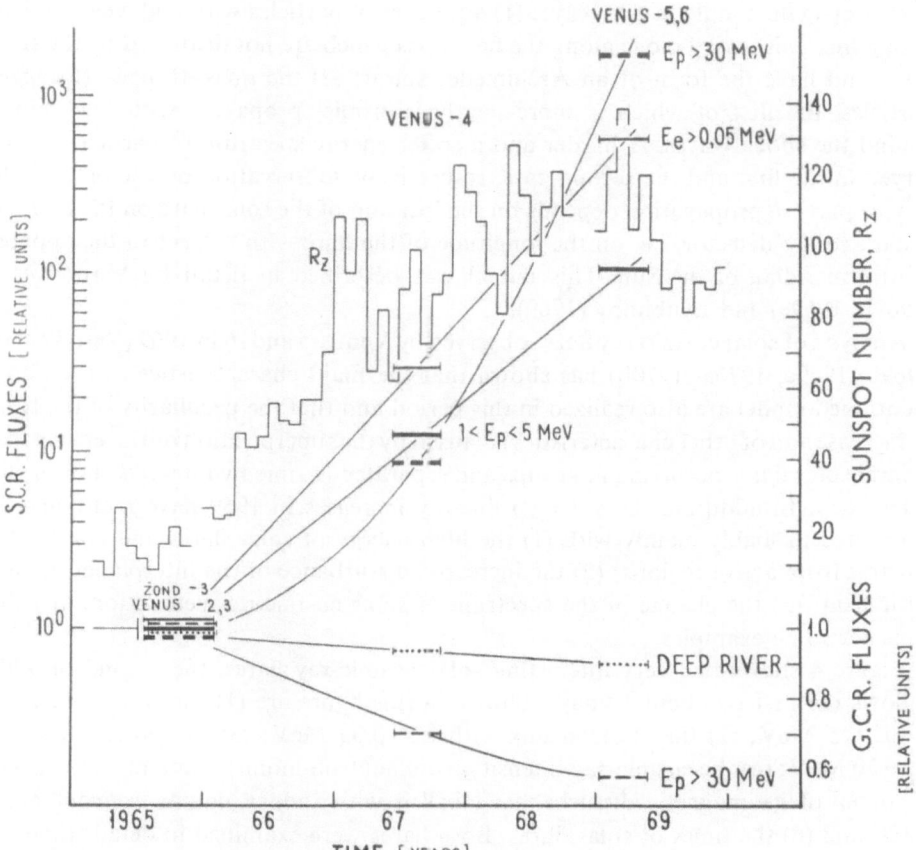

Fig. 3.   Time dependence of average solar cosmic-ray fluxes during space probe flights. The fluxes have been normalized to the mean values obtained during the 1965–1966 flights of Zond-3 and Venus-2 and -3 (1965–1966 = 1). The horizontal bars of thick broken lines correspond to the fluxes of the solar protons with $E_p > 30$ MeV. The thick solid bars show the proton fluxes with $1 < E_p < 5$ MeV. Thin solid bars show the electrons fluxes with $E_e > 0.05$ MeV. Thin dotted bars show the fluxes of galactic protons with $E_p > 30$ MeV. The light dotted line bars indicate the galactic cosmic-ray level according to the data from Deep River, with thin continuous lines connecting the disconnected pieces. The stepped line is the mean monthly sunspot number, R_z.

24% per AU. The values of the gradient obtained during various periods of the flight are different and vary from 55% to 16% per AU (Vernov and Logachev, 1969d). The value of the gradient and its changes may be due to either temporal or spatial causes. The result obtained will not be considered in detail here.

3. Solar Cosmic Ray Increases in 1969

Analysis of the low-energy solar cosmic-ray increases observed in 1965–67 has shown that after chromospheric flares on the Sun the low-energy cosmic rays produced by a flare propagate mainly in two ways: (1) a portion of particles with high velocities and strong flux anisotropy move along the field lines which are not disturbed by the shock wave and have the form of an Archimedes spiral; (2) the quasi-trapped (localized) particles, the flux of which is more nearly isotropic, propagate with low velocity behind the shock waves. A harder and a softer energy spectrum of particles are observed in the first and the second case, respectively. Observation of one or the other way of particle propagation depends on the location of the solar flare on the Sun with respect to the detector, i.e. on the longitude of the flare with respect to the apparent central meridian of the Sun. This model was described in detail by Vernov *et al.* (1969b, 1969e) and Lyubimov (1969).

Analysis of solar cosmic-ray flares observed by Venus-5 and -6 in 1969 (Vernov *et al.*, 1969b, 1969c, 1970a, 1970b) has shown that the main characteristics of the above-mentioned model are also realized in this period and that the peculiarity of the forms of increase and of other characteristics is caused by the superposition of the effects from several solar flares occurring in groups and separated in time by intervals shorter than a half day. In addition, the solar cosmic-ray increases in 1969 have many features connected probably mainly with (1) the high energy of solar flares and corpuscular streams from active regions; (2) the increased disturbance of the interplanetary magnetic field; (3) the change in the spectrum of solar cosmic-ray generation. We shall consider some examples.

Figure 4 shows two very interesting solar cosmic-ray flares, the second of which follows the first by about 12 days. Shown in this figure are (1) the proton flux with $1 < E_p < 5$ MeV; (2) the electron flux with $E_e > 0.05$ MeV; (3) the proton flux with $E_p > 30$ MeV; (4) the cosmic-ray intensity from neutron monitor data at Alert; (5) the K_p-index of geomagnetic disturbances labelled with sudden commencement marks (SC); and (6) the times of solar flares. Both flares were examined in detail; therefore we shall consider here only their main features. The solar flares (ESSA, 1965–69) occurred respectively on 29 March, 1969 at 19:23 (10°N, 54°E) and 10 April, 1969 at 04:14 (11°N, 90°E), that is, at the same latitude but at different longitudes on the Sun.

In the period under consideration during the Venus-5 and -6 flights, two groups of diametrically located active regions existed in the northern hemisphere of the Sun, which formed two broad but decidedly limited corpuscular streams (see Vernov *et al.*, 1969b, 1970d and Section 5 of the present report). The magnetic fields of these corpuscular streams, together with the shock-wave fields and the fields carried away by the flare, could form loop structures of the 'Gold's bottle' type (Gold, 1959).

The solar flares which occurred in these two groups of active regions in the eastern part of the Sun caused the shock waves behind the fronts and between the two corpuscular streams, the solar cosmic rays from these and subsequent flares were accumulated.

For the first flare, the time of increase in the proton flux with $E_p > 30$ MeV to the peak was 16 hr; for the second flare this time was 46 hr. The maximum solar cosmic-ray fluxes (in particles cm^{-2} sec^{-1} ster^{-1}) for the first and second flares were:

TABLE I

	30 March 1969	13 April 1969
$E_e > 0.05$ MeV	1.2×10^3	$\geqslant 10^5$
$1 < E_p < 5$ MeV	0.7×10^3	2.1×10^4
$E_p > 30$ MeV	13	1.8×10^2

The exponents γ of the integral energy spectra of the form $N(E) \propto E^{-\gamma}$ for the protons in the 1–5 MeV and the > 30 MeV energy ranges at the peaks of these flares were 1.2 and 1.4, respectively. These flares generated a great number of electrons. As can be seen from Figure 4 the accumulation of electron was very efficient.

In connection with the second flare (13 April, 1969), it is interesting to note that the flare of Class 3B occurring after it (on 21 April, 1969 at 20:30) at about the same point on the Sun (25°N, 33°W) did not cause any effect at either the Earth or Venus-6

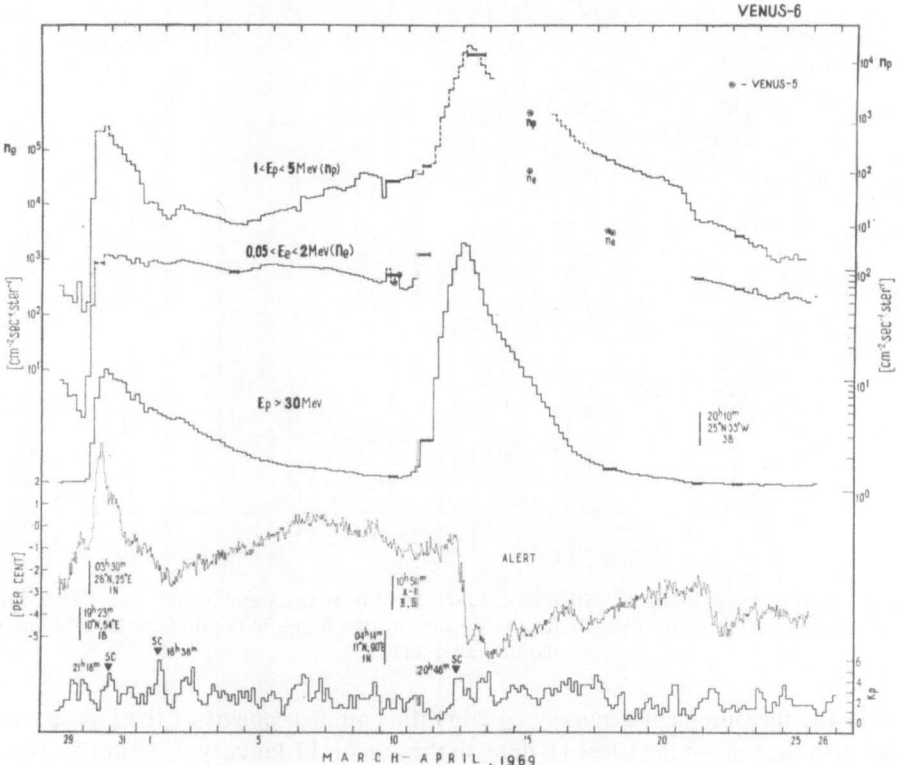

Fig. 4.   Two solar cosmic-ray (scr) bursts according to the data from Venus-5 and -6. Shown are the 4-hr values of the fluxes of (1) protons with $1 < E_p < 5$ MeV (labelled Np); (2) electrons with $0.05 < E_e < 2$ MeV (labelled Ne); (3) protons with $E_p > 30$ MeV. Dotted lines indicate unreliable values. Shown below for comparison are the hourly data from the neutron monitor at Alert and 3-hr data of the K_p-index labelled with sudden commencement (SC) marks. Vertical bars mark the moments of solar flares with notations on peaking time, coordinates, and class.

(Figure 4). It is possible that the effects of this flare were blocked by magnetic fields which were extended by the flares on 10 April, 1969, such as in the cases discussed by Gopasyuk and Krivsky (1967).

Consider now two cases of small flares of electrons with $E_e > 0.05$ MeV, for which the difference in propagation is attributed to the location (longitude) of the source. Figure 5 shows the electron fluxes for these flares. The first of them began on 17 January, 1969 at about 19:00. The electron flux increased slowly, for about 22 ± 4 hr, up to the maximum value 33 cm^{-2} sec^{-1} ster^{-1}. The duration of decrease was about 50 hr.

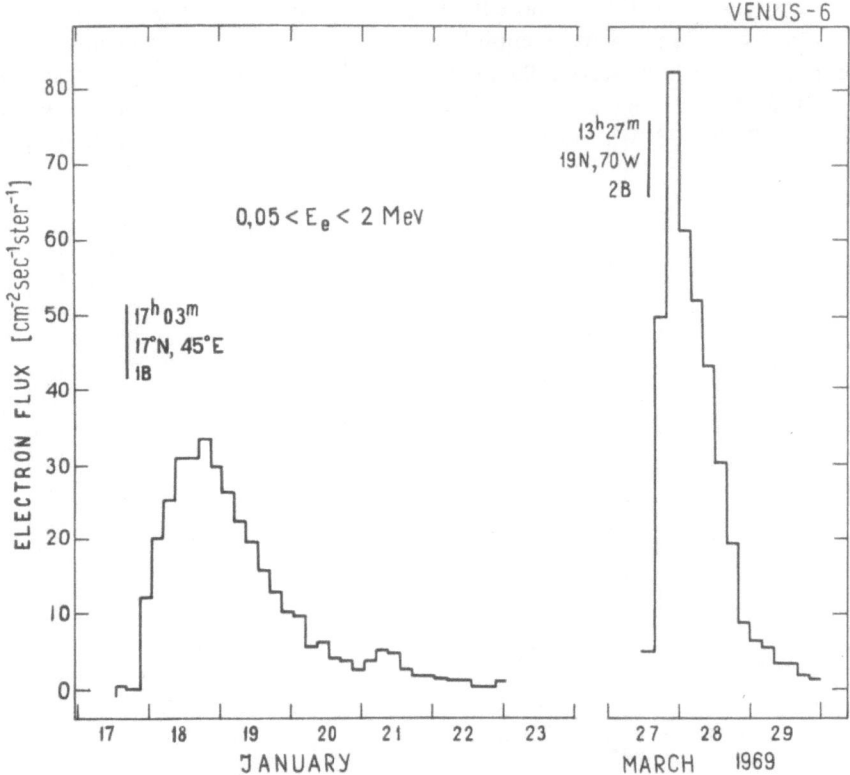

Fig. 5.   Two bursts of electrons with $0.05 < E_e < 2$ MeV in January and March, 1969, according to the Venus-6 data. The vertical bars mark the associated solar flares, w th notations on peaking time, coordinates, and class.

Within the measurement accuracy (± 2 hr) the commencement of the increase coincided with the time of the Class 1B flare on the Sun on 17 January, 1969 at 17:03 (peak) at 17°N, 45°E. The flare was accompanied by a powerful radio burst in the centimeter wavelength range. The radio-emission flux in the burst increased with frequency. At frequencies of 2800 MHz and 3500 MHz the maximum flux was 515×10^{-22} W m^{-2} Hz^{-1} and 2260×10^{-22} W m^{-2} Hz^{-1} respectively (ESSA, 1965–69). It may be considered that the above-mentioned flare and radio burst produced a great electron flux, the diffusion wave of which was observed on Venus-6. The main flux of electrons and protons, if these were generated, moved along spiral lines of force of the magnetic field, farther to the east of the space probe. Then, using the delay time of the flux peak with respect to the radio burst, we obtain the diffusion coefficient $\sim 4 \times 10^{22}$ cm^2 sec^{-1} and the mean diffusion velocity $V_e = r/t_m \approx 1700$ km/sec where r and t_m are respectively the distance and difference in time of the intensity peak of cosmic rays from the solar flare. The diffusion velocity is low and close to the shock-wave velocity. The commencement of a decrease in the intensity of galactic protons with $E_p > 30$ MeV and a depth of about 1% may be noted in the neighborhood of the peak. A small increase (of the order of 0.1 cm^{-2} sec^{-1} ster^{-1}) in the proton flux with $1 < E_p < 5$ MeV corresponds to this decrease (see Figure 8) (Vernov et al., 1969b). If one considers that all these events are interconnected, the observed electron flux may be explained by the transfer of electrons behind the shock wave and by the diffusion through the wave front during the transfer. Then the value obtained for the diffusion coefficient is meaningless.

The second electron flux increase began on 27 March, 1969 at about 16:00, which within measurement accuracy (± 2 hr) coincides with the time of the flare on the Sun. The electron flux reached the peak value of 83 cm^{-2} sec^{-1} ster^{-1} in 6 ± 4 hr; the duration of the decrease was about 30 hr.

The Class 2B solar flare which caused this electron increase occurred on 27 March, 1969 at 13:27 at 19° N, 70° W This flare, like the first one, was accompanied by radio bursts of Types II, III, and IV and an X-ray burst in the 1–8 Å range.

As can be seen from Table II below, the delay time of the electron peak is small and the increase in intensity is rapid when the flare occurs on the western side of the Sun;

TABLE II

Date and time of the flare	17 January, 1969 17:03	27 March, 1969 13:27
Coordinates	17°N 45°E	19°N 70°W
Duration of intensity increase	22 ± 4 hr	6 ± 4 hr
Duration of intensity decrease	50 hr	30 hr
Delay time of the intensity peak	24 ± 2 hr	8.5 ± 2 hr

but the delay of the peak is greater and the increase in flux is slower when the flare occurs on the eastern side. These two facts clearly illustrate the east-west asymmetry in low-energy electron propagation.

At the end of the description of some interesting cases of the solar cosmic ray increase caused by solar flares we shall present an example of the increase, related to an increase in the density and anisotropy of the background particles that had been isotropic earlier in front of the shock wavefront. This problem has been studied in a number of experimental works which considered acceleration of either galactic background particles (Dorman, 1963; Dorman et al., 1970) or the particles that remained from flares (Rao et al., 1967; Lanzerotti, 1969) at the shock wave-front. The left part of Figure 6 presents data for the event itself, whereas the right part shows the same parameters after an interval of 28–29 days, i.e. after one rotation of the Sun. Shown in this figure are: (1) anisotropy of the proton flux with $1 < E_p < 5$ MeV for the direction from and to the Sun, as shown by fluxes moving away from the Sun (thick line) and

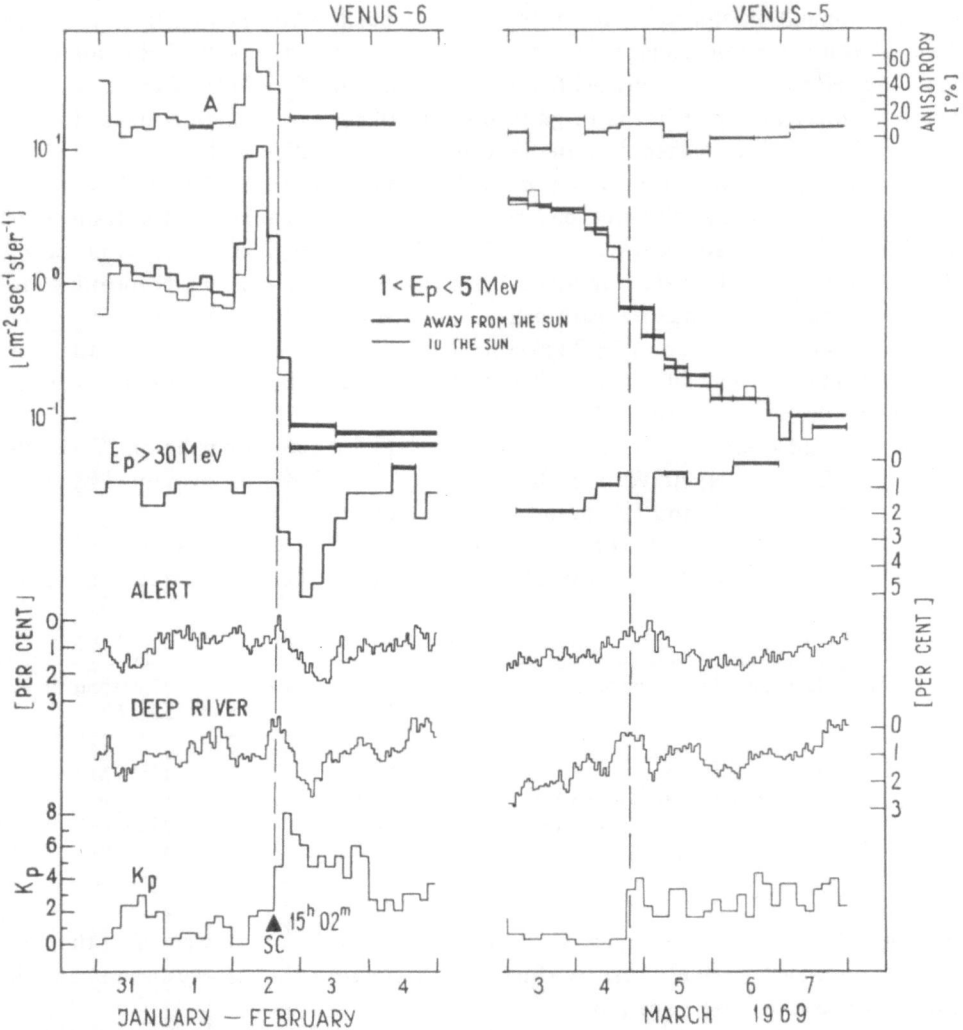

Fig. 6.   Increase in the proton flux with $1 < E_p < 5$ MeV at the front of the shock wave co-rotating with the Sun (Venus-5 and -6 data). The set of plots on the right side of the figure are the same as those on the left side, but for the next solar rotation. The thin vertical broken line marks the time when the effects are to be matched. The figure shows the 4-hr values of (1) the anisotropy $A = n^+ - n^-/n^+ + n^- \times \times 100\,\%$, where n^+ is the flux moving away from the Sun and n^- is that toward the Sun; (2) proton flux with $1 < E_p < 5$ MeV moving away from the Sun (thick line) and toward it (thin line); (3) proton fluxes with $E_p > 30$ MeV; (4) hourly values of neutron monitors at Alert and Deep River; (5) 3-hr values of the K_p-index labelled with the sudden commencement mark (SC).

toward the Sun (thin line); (2) intensity of galactic protons with $E_p > 30$ MeV; (3) neutron monitor data from Alert and Deep River; and (4) K_p-index of geomagnetic disturbance labelled with a sudden commencement mark (SC).

The isotropic proton fluxes with $1 < E_p < 5$ MeV which increased slowly in time and were observed on 31 January–1 February, 1969 and 3 March, 1969, were generated by the solar flares of 24 January and 24–27 February, 1969, respectively. These cases were described in detail by Vernov et al. (1969b, 1969c, 1970b).

We believe that a shock wave was observed at the Earth on 2 February, 1969 at 15:02, and on board Venus-5 and -6. On the Earth, this shock wave caused the sudden commencement of a magnetic storm (SC) (maximum value of $K_p = 8$) and a Forbush decrease of 2% depth and 0.15% per hour steepness. On Venus-5 and -6 the Forbush decrease had a depth of 4% and steepness of 0.4% per hour. This shock wave was a quasi-stationary wave corotating with the Sun (see McCracken et al., 1966; Davis, 1968). This wave appeared because of the presence of the contact surface where the quasi-stationary stream of hot gas met the solar wind. Such a stream might be caused by an active region which developed rapidly. During the next passage past the Earth and Venus-5 and -6, the velocity of the stream from the active region decreased and the shock wave weakened considerably or was transferred to the gap, as confirmed by the changes in the characteristics of this event after one solar rotation. (Similar events in Figure 6 were observed in the second half of 2 February, 1969, and late on 4 March, 1969).

The peak of the proton flux with $1 < E_g < 5$ MeV and the anisotropy of 67% observed on Venus-5 and -6 on 2 February, 1969 were due to the reflection of the background isotropic particles from the shock wave approaching them and to their 'sliding' along the front of this wave in the direction away from the Sun. During the next passage the wave, which had weakened considerably, did not cause particle acceleration; only a Forbush decrease in the particle intensity was observed (Figure 6). The observed effects are likely to be connected with the boundary of the interplanetary magnetic field sector.

4. Characteristics and Forbush Decreases in Galactic Cosmic Rays in 1969

It has been shown in Section 2 of this paper that in 1969, during the Venus-5 and -6 flights, Forbush decreases were more numerous than in 1965–67. However, only three cases could be selected for the comparative qualitative analysis of the characteristics of Forbush decreases of galactic cosmic rays with energies > 30 MeV: on 23 March, 27 April, and 14 May, 1969. Other Forbush decreases were partly or completely 'filled' with solar protons of $E_p > 30$ MeV. In most cases the mean velocities of the shock waves and the gas were difficult to determine from the delay time of the commencement and minimum of Forbush decreases (Vernov et al., 1969b; Lyubimov, 1968) because of ambiguous connection of these events with solar flares.

The three above-mentioned Forbush decreases, which were sufficiently clearcut, were examined in detail by Vernov et al. (1970c); Vernov et al. (1970a) made an at-

tempt to examine the effects of 30 March, 1969 and 12 April, 1969; again Vernov *et al.*
(1969b, 1969c, 1970b) presented the other less well defined cases.

Here we shall show only a single example of Forbush decrease with anomalously
rapid recovery of the galactic cosmic ray intensity. Figure 7 shows (1) four-hour values
of the intensity (in per cent change) of protons with $E_p > 30$ MeV, obtained from
Venus-6; (2) hourly values of the intensity (in per cent change) of the neutron com-

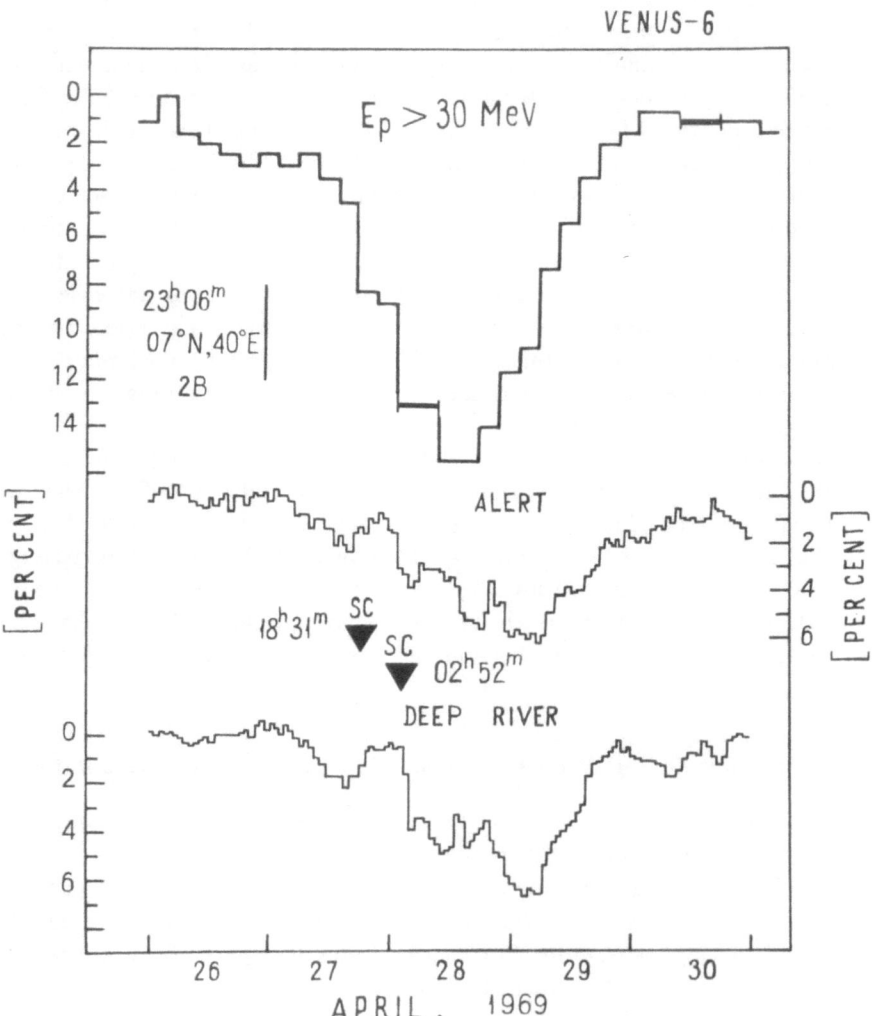

Fig. 7.　Forbush decrease on 27–29 April 1969. Shown in the figure are (1) 4-hr values of the 30-MeV
proton intensity, according to the Venus-6 data; and (2) hourly values of the neutron monitors at
Alert and Deep River. Also, (3) the vertical bar shows the moment of the solar flare with notes on the
peaking time, coordinates, and class, and (4) the black triangles labelled SC show the times of
sudden commencements.

ponent, according to the data from stations at Alert and Deep River; and (3) time marks for the solar flare and two sudden commencements (SC).

On 26 April, 1969 at 23:06, a Class 2B (2N) flare at 07° N, 40° E, reached peak intensity. On 27 April, 1969 at about 10 hr a Forbush decrease began at Venus-6. In 8 hr the steepness of the decrease increased abruptly. The full depth of the decrease was 13%, at a mean steepness of 0.52% per hour. The steepness of recovery was 0.67%. Durations of the decrease and recovery were about the same (25–30 hr).

The Forbush decrease at the Earth, as at Venus-6, had a double fall. The total depth was 4%, with a steepness of 0.16% per hour. The recovery rate was 0.21% per hour. Durations of decrease and recovery were also about the same (25–30 hr). Two sudden commencements were observed on Earth: on 27 April, 1969 at 18:31, and on 28 April, 1969 at 02:52. If the first and second sudden commencements are compared with the Forbush decrease commencement and the abrupt increase in the steepness of the intensity decrease on Venus-6 respectively, the mean velocities of the movement, calculated from the delay between the flare peak and three instants in time for Venus-6 and the Earth, will be $V_1(I) = 2930^{+650}_{-450}$ and $V_2(I) = 2140$ km/sec; $V_1(II) = 1900^{+250}_{-200}$ km/sec and $V_2(II) = 1490$ km/sec, and the velocities between the Venus-6 and Earth will be $V_3(I) = 1130^{+340}_{-220}$ km/sec and $V_3(II) = 870^{+200}_{-130}$ km/sec. The difference of heliocentric distances between Venus-6 and the Earth was 34.5×10^6 km and Venus-6 was 11° west of the Earth in longitude.

In accordance with the case examined (Hirshberg et al., 1970) the first and second sudden commencements may be considered to be the moments of arrival of the shock wave and of the gas piston with a helium layer, respectively. An abrupt increase in the plasma flux corresponds to the moment of the steep intensity decrease on Venus-6 (Gringauz, priv. comm.). The ratios of the mean velocities of the shock wave and gas show a strong slowing of the motion.

The rapid recovery of the galactic cosmic-ray intensity may be explained by several causes. One may suppose the burst character of the flare (strong slowing) (Hundhausen and Gentry, 1969) and, in connection with this, a thin transparent layer of magnetised gas. Besides that, during, before, and after the Forbush decrease the Deep River neutron monitor detected a large one-day intensity variation with a peak at about 18 hr local time, i.e. the shock wave and gas layer were oriented approximately in the direction of undisturbed interplanetary field and did not hamper the galactic cosmic-ray penetration to the Sun. Another cause of the rapid recovery of the galactic cosmic-ray intensity consists in the probable fact that, as the analysis of the active region location on the Sun shows (Vernov et al., 1970d) this region of the space corresponded to the 'gap' between two groups of active regions on the Sun and hence to the relatively 'empty' space between powerful corpuscular streams from active regions.

All the causes of unusually rapid recovery of the galactic cosmic-ray intensity supplement each other; only accurate measurements of the magnetic field and plasma however, can finally confirm this conception.

Consideration of the dynamic characteristics of the flare disturbances in the interplanetary medium by means of the study of Forbush decreases in galactic cosmic rays at various distances from the Sun has enabled us to draw some new conclusions.

Ratios of the mean velocities of shock waves calculated at various distances from the Sun from the delay of Forbush decrease commencements, or sudden commencements with respect to the solar flare peak, show both the deceleration and acceleration of the waves. This method of calculation gives the correct result only in the case of (1) a shock front orthogonal to the detector-Sun line, (2) a sufficiently short period of plasma injection from the flare region.

Proceeding from the necessity of deceleration of shock waves if spherical, (with the center on the Sun, if they are spherical) (Landau and Lifshits, 1953; Hundhausen and Gentry, preprint) and considering the considerable deviations of the wave-front from orthogonality to the Earth-Sun line (Hirshberg, 1968; Wilcox, 1969; Hirshberg et al., 1970), it appears that the 'acceleration' effect obtained indicates a strong directiveness in flare ejections and shock waves, and a prolonged injection of hot gas after the flare (Hirshberg et al., 1970). It will be noted that the steepness of the recovery after Forbush decrease minimum may be also connected with duration of the flare gas ejection (Vernov et al., 1970c).

The following changes in flare disturbances took place from 1965–1967 to 1969: (1) the mean velocity of the shock waves increased, but not much; and (2) the period of the flare gas ejection increased (the value of the deceleration of mean velocities decreased). Thus, the increase in the flare energy from 1965–1969 is probably due to an increase in the initial velocities of the ejection. It may be supposed that durable flares are of stronger directivity and that the shock-wave damping is due not only to energy but also to the power (based on duration and initial velocity). This is, at the same energies more powerful flares, i.e. those shorter in duration but having a higher initial velocity, are damped more rapidly.

It should also be noted that all Forbush decreases are probably an indication of shock waves pushed by a gas piston. Statistical analysis of the Forbush decrease parameters and magnetic field values (as measured in 1967 from Venus-4) has shown (Gorchakov et al., 1970) that the magnetic field strength increases from the commencement to the minimum of the Forbush decrease and then falls rapidly. This result corresponds to the magnetic field distribution between the shock wave and gas piston obtained by Hirshberg et al., 1970).

5. Quasistationary Forbush Decreases of Galactis Cosmic Rays and Active Regions on the Sun.

Vernov et al. (1968b, 1969e) present a comparison of active regions on the Sun with quasistationary Forbush decreases of galactic cosmic rays using results obtained in 1965–1966. This method made it possible to determine the large-scale structure of magnetic fields of quasistationary corpuscular streams and to connect this structure with active regions. It has been shown that the parameters of such Forbush decreases are determined by the dynamics of active regions, their latitude, and phase of development. The results obtained agree with the conclusions drawn from a comparison of direct measurements of interplanetary and photospheric magnetic fields (Schatten et al., 1968, 1969).

A good correspondence between the 'power' of an active region (product of brightness and flocculus area) and Forbush decreases depth, and a weak latitude effect were obtained in 1965–1966. In 1967 a larger latitude effect was obtained. It was shown that the Forbush decreases connected with one group of active regions were 'filled' with solar protons with $E_p > 30$ MeV.

According to the 1969 analysis (Vernov et al., 1970d) the solar active regions were grouped in two diametrically opposite locations (at about 50–130° and 205–340° of Carrington longitude). In accordance with such a location of active regions, two persistent 'tray-like' sequences of Forbush decreases, with steep decrease and recovery, and a section of plateau at the normal level of galactic cosmic rays between them, corresponding to the 'gap' between active region groups, were observed on Venus-5 and -6 and, also, according to the neutron monitor data, on the Earth. A strong latitude effect was observed.

In different years the latitude dependences obtained probably correspond to the coronal ray distribution at various phases of solar activity.

6. Conclusions

Concluding the analysis of main results of cosmic ray studies in 1969 and comparing them with those obtained earlier, one should note the great variety of the events observed. It has been shown as a result that, from 1965–1969:

(1) Solar proton fluxes with $1 < E_p < 5$ MeV increased by a factor of about 110. The increase may be approximated by an exponential function with a time constant of 0.7 yr.

(2) Solar electron fluxes with $E_e > 0.05$ MeV increased by a factor of about 580. The increase may be approximated by an exponential function with a time constant of 0.5 yr.

(3) Solar proton fluxes with $E_p > 30$ MeV increased by a factor of about 1700. The rise from 1967–1969 is abrupt.

(4) The solar proton spectrum in the 1–30 MeV energy range becomes harder.

(5) The level of galactic protons with $E_p > 30$ MeV decreased by a factor of 1.7. The decrease was slower from 1967–1969.

(6) Mean velocities of shock waves and Forbush decrease amplitudes increased approximately linearly in time.

(7) The latitude effect of solar active regions on the galactic cosmic-ray flux increased.

The following effects were observed in 1969:

(8) Accumulation of solar cosmic rays from powerful flares was observed at a distance of 1 AU from the Sun.

(9) Protons with $E_p > 30$ MeV were effectively contained between the shock wave front and the front of the flare plasma.

(10) Electron bursts at $E_e > 0.05$ MeV show a longitude effect.

(11) An example of acceleration of protons with $1 < E_p < 5$ MeV was observed at the front of a shock wave co-rotating with the Sun.

(12) The class 3B flare of 21 April, 1969 did not produce any observable effects in the interplanetary medium at a distance of 1 AU from the Sun.

(13) Persistent pushing by flare gases was observed.

(14) Directionality of solar effects from powerful flares was observed.

References

Davis, L.: 1968, *Solar Wind*, edit. Mir. Moscow, p. 185.

Dorman, L. I.: 1963, *Cosmic Ray Variations and Space Exploration*, ed. Acad. Sci., U.S.S.R.

Dorman, L. I., Kaminer, N. S., and Kebuladse, T. V.: 1970, 'Galactic Cosmic Ray Acceleration by Strong Shock Waves in the Interplanetary Space', report at Internat. STP Symp., Leningrad.

ESSA, 1965–1969: IER-FB Series (Solar-Geophysical Data), ESSA Laboratories, Boulder, Colo., U.S.A [now NOAA Environmental Research Labs.].

Gold, T.: 1959, *J. Geophys. Res.* **64**, 1665.

Gopasyuk, S. and Krivsky, L.: 1967, *BAC* **18**, 135.

Gorchakov, E. V., Ignatyev, P. P., and Galachyev, N. G.: 1970, 'On the Galactic Cosmic Ray Interaction with Magnetic Fields', *Space Res.* in press.

Gringauz, K. I.: private communication.

Hirshberg, J.: 1968, *Planetary Space Sci.* **16**, 309.

Hirshberg, J., Alksne, A., Colburn, D. S., Bame, S. J., and Hundhausen, A. J.: 1970, 'Observation of a Solar Flare Induced by Interplanetary Shock and Helium Enriched Driver Gas', *J. Geophys. Res.* **75**, 1.

Hundhausen, A. J. and Gentry, R. A.: 1969, 'Numerical Simulation of Flare-Generated Disturbances in the Solar Wind', A.S.L. of Univ. Calif., *J. Geophys. Res.* **74**, 2908.

Landau, L. D. and Lifshits, E. M.: 1953, *Solid Medium Mechanics*, Gostekhisdat.

Lanzerotti, L. J.: 1969, *J. Geophys. Res.* **74**, 2851.

Lyubimov, G. P.: 1968, *Astron. Circular*, Acad, Sci., U.S.S.R., No. 488, 4.

Lyubimov, G. P.: 1969, Theses, NIIYaF MGU, Moscow.

McCracken, K. G., Rao, U. R., and Bukata, R. P.: 1966, *Phys. Rev. Letters* **17**, 928.

Rao, U. R., McCracken, K. G., and Bukata, R. P., 1967, *J. Geophys. Res.* **72**, 4325.

Schatten, K. H., Ness, N. F., and Wilcox, J. M.: 1968, *Solar Phys.* **5**, 240.

Schatten, K. H., Wilcox, J. M., and Ness, N. F., 1969, *Solar Phys.* **6**, 442.

Vernov, S. N., Chudakov, A. E., Vakulov, P. V., Logachev, Yu. I., Lyubimov, G. P., and Peresligina, N. V.: 1967, *Izv. Akad. Nauk*, S.S.R, ser. fis., **31**, No. 8, 1255.

Vernov, S. N., Chudakov, A. E., Vakulov, P. V., Gorchakov, E. V., Kontor, N. N., Kusnetsov, S. N., Logachev, Yu. I., Lyubimov, G. P., Nikolaev, A. G., Pereslegina, N. V., and Tverskoy, B. A.: 1968a, Proc. 5th All-Union Winter School on Cosmophysics, Apatity, p. 5.

Vernov, S. N., Logachev, Yu. I., Lyubimov, G. P., and Pereslegina, N. V.: 1968b, 'Galactic Cosmic Ray Variations and Quasistationary Solar Plasma Fluxes in Space', Rep. at the All-Union Conf. on Cosmic Ray Physics, Tashkent; Rep. at Internat. Symp. on STP, Crimean Astrophysical Obs., Acad. Sci., U.S.S.R.

Vernov, S. N., Chudakov, A. E., Vakulov, P. V., Gorchakov, E. V., Ignatyev, P. P., Kontor, N. N., Kusnetsov, S. N., Logachev, Yu. I., Lyubimov, G. P., Nikolaev, A. G., and Pereslegina, N. V.: 1969a, *Space Res.* **9**, Report to COSPAR, Tokyo, 1968.

Vernov, S. N., Chudakov, A. E., Vakulov, P. V., Gorchakov, E. V., Kontor, N. N., Logachev, Yu. I., Lyubimov, G. P., Pereslegina, N. V., and Timofeev, G. A.: 1969b, Report at Internat. Conf. on Cosmic Ray Physics, Budapest.

Vernov, S. N., Kontor, N. N., Lyubimov, G. P., Pereslegina, N. V., and Chuchkov, E. A.: 1969c, Report at Internat. Symp. on Satellite Measurements, Holland.

Vernov, S. N. and Logachev, Yu. I.: 1969d, 'Gradient of the Galactic Cosmic Rays with Energy $E_p > 30$ MeV During January-May 1969', in press.

Vernov, S. N., Lyubimov, G. P., and Pereslegina, N. V.: 1969e, Report at 6th Summer School of Cosmophysicists, Apatity.

Vernov, S. N., Chuchkov, E. A., Kontor, N. N., Lyubimov, G. P., and Pereslegina, N. V.: 1970a, 'Solar Cosmic Ray Flares with Proton Energy $E_p > 30$ MeV According to the Measurements from Venus-6 Space Probe in March-April 1969', report to COSPAR, Leningrad.

Vernov, S. N., Kontor, N. N., Lyubimov, G. P., Pereslegina, N. V., and Chuchkov, E. A.: 1970b, 'Analysis of Solar Cosmic Ray Increases According to the Measurements from Venus-5, 6 Space Probes', report to COSPAR, Leningrad.

Vernov, S. N., Lyubimov, G. P., Kontor, N. N., Pereslegina, N. V., and Chuchkov, E. A.: 1970c, 'Dynamical Characteristics of Forbush Decreases of galactic Cosmic Rays in the Interplanetary Space from January to May 1969', report to Internat. STP Symp. Leningrad.

Vernov, S. N., Pereslegina, N. V., Kontor, N. N., Lyubimov, G. P., and Chuchkov, E. A.: 1970d, 'Confrontation of Solar Active Regions with Galactic Cosmic Ray Variations from January to May 1969', report to Internat. STP Symp. Leningrad.

Wilcox, J. M.: 1969, *Solar Flares and Space Research*, North Holland Publ. Co., Amsterdam, p. 294.

ENERGETIC SOLAR PARTICLES
IN THE INTERPLANETARY MEDIUM*

W. I. AXFORD

Dept. of Physics, Dept. of Applied Physics and Information Science, University of Calif., San Diego
La Jolla, Calif., 92037

Abstract. Models which describe the diffusion of energetic solar particles in the interplanetary medium are reviewed. Most attention is paid to the case of impulsive emission from solar flares, taking into account the effects of the interplanetary magnetic field, diffusion near the Sun, convection, and energy losses associated with the expansion of the solar wind. The case of continuous emission is also considered briefly.

1. Introduction

In this review we discuss the various models that have been developed to describe the behavior of energetic particles emitted by the Sun. The emission can occur in the form of well-defined 'events' following solar flares, or take place on a more or less continuous basis (perhaps as a result of the superposition of many small flares). From the theoretical point of view most attention has been paid to 'events' and consequently the major part of this review is devoted to theories describing the behavior of particle fluxes released impulsively at the Sun. However the importance of continual emission at low energies ($\lesssim 50$ MeV/nucleon) has recently become appreciated, and some theoretical work has been carried out.

The characteristic features of energetic particle events are as follows:

(1) The mean intensity varies with time in a manner which indicates that the particles do not propagate directly from their region of origin to the Earth. This suggests that the particles might be stored near the Sun for some time, and that scattering in the interplanetary medium is important. The intensity often appears to decay exponentially with a time constant of the order of 10–20 hr (e.g. Anderson, 1969).

(2) Flares occurring on the western visible hemisphere of the Sun generally give rise to larger events at the Earth, with a prompt arrival of particles. Events resulting from flares in the eastern hemisphere tend to be smaller and to increase more slowly to maximum intensity (e.g. Fichtel and McDonald, 1967).

(3) The anisotropy is initially aligned with the interplanetary magnetic field, being in the direction away from the Sun following the magnetic field lines outwards. The initial anisotropy is usually large ($\sim 100\%$) for west-side flares, but is less pronounced (~ 10–50%) for east-side flares (e.g. McCracken, 1963; McCracken *et al.*, 1967).

(4) The anisotropy decays with time, which is again consistent with scattering in the interplanetary medium. At high energies the anisotropy decays to a low value ($\lesssim 10\%$), so that the intensity observed by neutron monitors for example is more or less isotropic

* This work was supported by the National Aeronautics and Space Administration under Contract NGR-05-009-081.

late in the event. At low energies ($\lesssim 50$ MeV/nucleon), the anisotropy decays to a constant 'equilibrium' value and the direction changes to radial from the Sun (e.g. McCracken *et al.*, 1967, 1968; Rao *et al.*, 1969).

(5) The spectrum, when referred back to the source, is soft (i.e. the spectral index is typically 3–6 if a power-law fit is used). The composition is apparently similar to that of the solar photosphere as far as heavier species with respect to helium are concerned. Both the spectrum and the composition appear to be distorted by propagation effects (Biswas and Fichtel, 1965; Fichtel and McDonald, 1967).

(6) Disturbances in the solar wind can have a pronounced effect on the behavior of low energy particle fluxes, giving rise to pre-sudden commencement increases ('spikes') associated with interplanetary shocks (e.g. Armstrong *et al.*, 1970; Ogilvie and Arens, 1971), to energetic storm particle events (Anderson, 1969), and (occasionally) to a complete disappearance of the particle fluxes (e.g. Axford and Reid, 1962).

The nature of particle fluxes which are emitted from the Sun on a more or less continuous basis is less well-established, since it is often difficult to decide whether particles being observed at any given time are of galactic or solar origin. However, it is probably correct to characterize them as follows:

(1) The particles can be distinguished from galactic cosmic rays only at low energies ($\lesssim 50$ MeV/nucleon), and are probably always dominant at the lowest energies (e.g. Kinsey, 1970).

(2) The composition is probably similar to that of the photosphere except that nuclear interactions may be expected to produce observable quantities of secondary particles (e.g. deuterons, tritons, He^3 nuclei, positrons etc.) since the storage time in the solar corona could be quite long (~ 5–10 days). The spectrum may be affected by ionization losses as well as by propagation effects.

(3) The mean (omni-directional) intensity of the particle distribution should have a pronounced radial gradient ($\gtrsim 200\%$ AU), decreasing outwards from the Sun (e.g. Krimigis and Venkatesan, 1969; Vernov *et al.*, 1969).

(4) The anisotropy should also be outwards from the Sun, and may be rather similar to the 'equilibrium' anisotropy observed at low energies in the final phase of solar events (e.g. Gleeson *et al.*, 1970).

(5) Slow time variations may be apparent as the sources evolve. Since the sources are presumably associated with 'active regions' on the Sun, 27-day periodicities must be evident (e.g. Desai and McDonald, 1970).

Models of energetic particle events are described in Section 2, 3 and 4. Isotropic diffusion theory is discussed in Section 2, anisotropic diffusion theory is discussed in Section 3, and the effects of convection and energy loss in Section 4. Models describing the behavior of energetic particles emitted continuously by the Sun are discussed in Section 5. We have not attempted to review the theory of particle scattering in an irregular magnetic field, but refer the interested reader to the work of Jokipii (1966, 1967, 1968a, b), Roelof (1966, 1968), Jokipii and Parker (1969), Hasselmann and Wibberenz (1968), Dolginov and Toptygin (1968), Klimas and Sandri (1970), and Hall and Sturrock (1967).

2. Isotropic Diffusion

The first and simplest models proposed for the explanation of solar energetic particle events were based on the concept of isotropic diffusion of particles in the interplanetary medium, assuming spherical symmetry, and that energy changes and convection by the solar wind can be neglected. In these models it is assumed in effect that a group of particles in a given kinetic energy range behaves like a gas which seeps through a scattering background at a speed which is small compared with the particle speed (cf. Axford, 1965a). The equations describing the behavior of such a gas are

$$\frac{\partial U}{\partial t} + \frac{1}{r^2} \frac{\partial}{\partial r} (r^2 S) = 0, \tag{1}$$

$$\frac{\partial p}{\partial r} = -\frac{mS}{\tau}, \tag{2}$$

where $U(T)$ and $S(T)$ are the mean density and radial current density of particles in the kinetic energy range $(T, T+\mathrm{d}T)$, m is the particle mass, τ the 'relaxation' time, and r and t the heliocentric radial distance and time respectively. The partial pressure of the gas is taken to be

$$p = \tfrac{1}{3}mv^2 U, \tag{3}$$

where v is the particle speed ($v^2 = 2T/m$ for non-relativistic particles). This assumption for p, together with the neglect of the inertia terms in (2), is adequate provided $S \ll vU$ (e.g. Fisk and Axford, 1969a). The effects of convection by the solar wind can be taken into account by substituting $(S - VU)$ for S on the right hand side of (2), with V being the solar wind speed. However if $S \gg VU$ convection can be neglected, and in any case there is no point in allowing for convection if the effects of energy changes are not also taken into account (see Section 4).

The radial current density can be eliminated from these equations to yield the familiar diffusion equation for U:

$$\frac{\partial U}{\partial t} = \frac{1}{r^2} \frac{\partial}{\partial r} \left(r^2 \kappa \frac{\partial U}{\partial r} \right), \tag{4}$$

where the diffusion coefficient $\kappa \, (= \tfrac{1}{3}v^2\tau)$, is in general a function of r and T. τ can be considered as the mean 'collision' time for the particles as they interact with the irregular component of the interplanetary magnetic field; it can be determined independently from observations of fluctuations in the magnetic field (e.g. Jokipii and Coleman, 1968).

It is important to note that although the theoretical problem has been reduced to one of solving a familiar equation for U, the radial current density S plays just as important a role in the theory as U. Thus one cannot claim success for a particular model on the basis of comparisons of the observations with theoretical predictions of U only, since S is specifically predicted by any model and should also be included in such comparisons. U and S are related to the mean (omnidirectional) intensity, j_0, and the radial anisotropy ξ_r, as follows:

$$j_0 = vU/4\pi, \quad \xi_r = 3S/vU \tag{5}$$

(cf. Gleeson and Axford, 1967).

The development of the isotropic diffusion model for solar energetic particle events has been chiefly due to Parker (see Parker, 1963, for a complete summary). It is assumed that the particles are released impulsively at the Sun ($r=0$) at time $t=0$ so that the initial distribution is given by $U = N\delta(r)$, where N is a constant and $\delta(r)$ is the delta function. The various models differ only in the radial dependence assumed for the diffusion coefficient.

If we assume that $\kappa = \kappa_0(T)r^\beta$, then*

$$U(r, t) = \frac{N \exp\{- r^{(2-\beta)}/(2 - \beta)^2 \kappa_0 t\}}{4\pi\Gamma\left(\dfrac{3}{2 - \beta}\right)(2 - \beta)^{(4+\beta)/(2-\beta)}(\kappa_0 t)^{3/(2-\beta)}}; \tag{6}$$

thus the time of maximum intensity is given by

$$t_{max} = r^{2-\beta}/\{3(2 - \beta)\kappa_0\}, \tag{7}$$

and for large times ($t \gg t_{max}$) the intensity decays as $t^{-3/(2-\beta)}$ (Parker, 1963). There are three independent parameters which can be used to fit the observations, namely N, κ_0 and β, however N is only a scaling parameter. κ_0 and β can be determined by noting that a plot of $\log_e(j_0 t^{3/(2-\beta)})$ against $1/t$ should be a straight line if β is chosen correctly, and that the slope of the line (or alternatively t_{max}) determines κ_0 once β is known. Krimigis (1965) has made extensive use of this model, and has been able to match the data on the mean intensity in a number of events with reasonable success (see Figure 1). It is usually found that β is of the order of unity, and the 'mean free path' λ ($=3\kappa/v$) is of the order of 0.1 AU at the Earth's orbit for protons in the kinetic energy range 50–500 MeV. The anisotropy is radial and given by

$$\xi_r = \left(\frac{3}{2 - \beta}\right)\frac{r}{vt}, \tag{8}$$

which is independent of the diffusion coefficient (Dorman, 1963; Axford, 1965b). The time-scale for the decay of the anisotropy varies from ~ 30 min for relativistic particles to ~ 5 hr for 10 MeV protons. There have been very few attempts to match predictions of the anisotropy with observations (see Figure 2; also Quenby et al., 1969), and in some cases it is clear that no fit can be achieved even when the model adequately fits observations of the mean intensity (e.g. McCracken et al., 1967).

* It is evident from similarity arguments that the solution must be of the form $U = N(\kappa_0 t)^{-3/(2-\beta)}g(\theta)$, where $\theta = r^{(2-\beta)}/\kappa_0 t$ and $g(\theta)$ satisfies the ordinary differential equation $(2-\beta)^2[(2-\beta)\theta g'' + 3g'] + [(2-\beta)\theta g' + 3g] = 0$. The solution of this equation which is applicable in this case is $g(\theta) = A \exp\{-\theta/(2-\beta)^2\}$, the constant A being determined from the condition $\int_0^\infty 4\pi r^2 U \, dr = N$.

To obtain an exponential rather than a power-law decay of the mean intensity with time it is necessary to change either the nature of the source or the radial dependence of the diffusion coefficient. The simplest model involves an impulsive source with $\kappa = \kappa_0 \, r^\beta$ in $r<a$ where $a>1$ AU, and with free escape of particles at $r=a$ (i.e. $U=0$

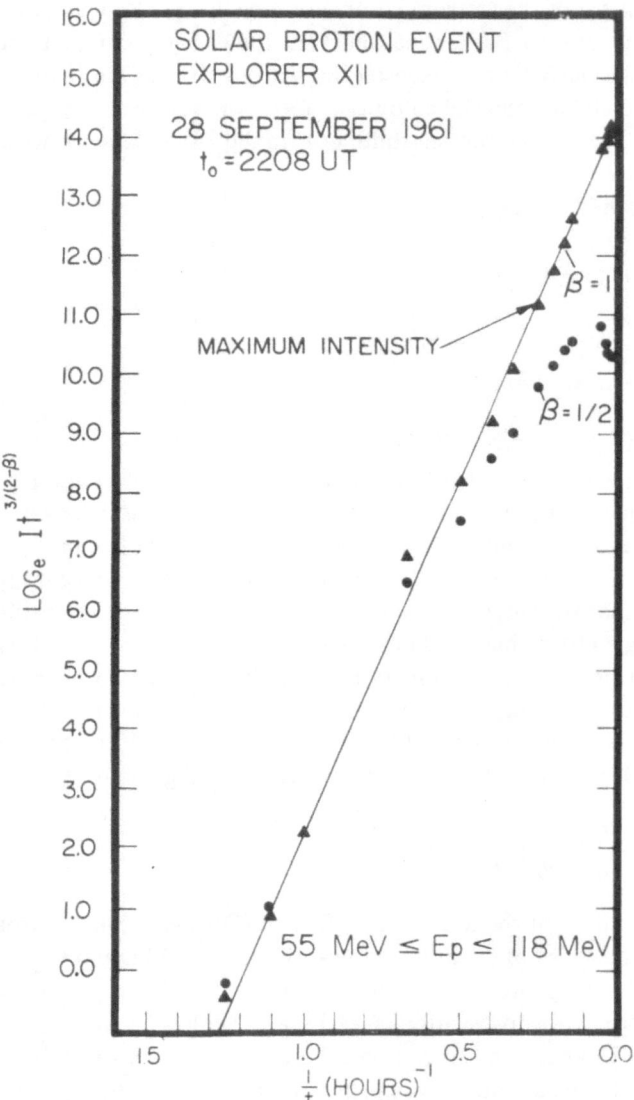

Fig. 1.   Data obtained during the event of 28 September 1961, by Bryant *et al.* (1965), plotted accor-ding to the model described by Equation (6) (from Krimigis, 1965). *I* is the same as the intensity j_0 of Equation (5). The data points are well-organized into a straight line by the choice $\beta=1$ for more than 20 hr after the flare. The corresponding value of the diffusion coefficient obtained from the slope of the line is $\sim 4 \times 10^{22}$ cm^2 sec^{-1}.

at $r=a$). For the case $\beta=0$, the solution of (4) is (see Parker, 1963)

$$U(r, t) = \frac{N}{2a^2 r} \sum_{n=1}^{n=\infty} n \sin\left(\frac{n\pi r}{a}\right) \exp(-n^2\pi^2\kappa_0 t/a^2). \tag{9}$$

The small time approximation $(a^2 \gg \kappa_0 t \gg r^2)$ is essentially the same as (6) with $\beta=0$, and thus the observations early in the event can be used to determine κ_0 from a plot similar to those shown in Figure 1. The large time approximation $(t \gg a^2/\kappa_0)$ is simply

$$U(r, t) \sim \frac{N}{2a^2 r} \sin\frac{\pi r}{a} \exp(-\pi^2\kappa_0 t/a^2), \tag{10}$$

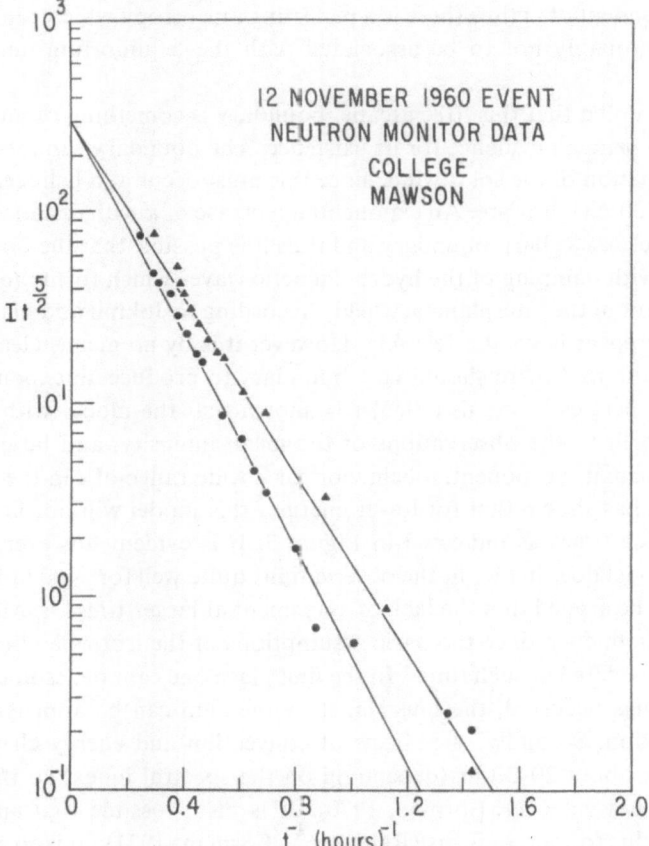

Fig. 2. Neutron monitor data from College (triangles) and Mawson (dots) for the event of 12 November 1960 (from Burlaga, 1967). The College neutron monitor was looking approximately along the local interplanetary magnetic field direction, while the neutron monitor at Mawson looked in the opposite direction. I is proportional to the neutron monitor counting rates and therefore to the directional intensity. According to the model in which $\kappa = \kappa_0 r^\beta$, the two sets of data should be organized into two straight lines in this plot if β is chosen correctly. The difference in the slopes of the lines should be approximately $2\xi t = 6r/(2-\beta)v$ (Fisk and Axford, 1969a). With $\beta = \frac{4}{3}$ the predicted slope difference is 60–70 min, while the observed slope difference is about 63 min, which is surprisingly good agreement.

and thus from the decay phase of the event it is a simple matter to determine $\pi^2 \kappa_0 / a^2$ and hence a. There are three independent parameters (four if β is not chosen to be zero), with N being again simply a scale factor. This type of model has been used by Hofmann and Winckler (1963) and by Bryant et al. (1962) to analyze their data, and they typically find that the scattering mean free path is of the order of 10^{-2}–10^{-1} AU, and the heliocentric distance to the 'boundary' is about 2 AU. For this model the anisotropy is again radial, and given by (8) for small times. During the decay phase however

$$\xi_r \sim \frac{3\kappa_0}{av} \left[\frac{a}{r} - \pi \cot \frac{\pi r}{a} \right], \tag{11}$$

which is independent of t; thus there is a persisting anisotropy which can be large near $r = a$. This is probably not to be associated with the 'equilibrium' anisotropy (see Section 4).

It should be noted that this 'free escape' boundary is something of an artifice since there is no independent evidence for its existence. The boundary cannot be associated with the termination of the solar wind, since this must occur at a heliocentric distance of the order of 30 AU or more. An exponential increase of κ with distance would serve the same purpose as a sharp boundary and thus it is possible that the boundary could be associated with damping of the hydromagnetic waves which form most of the irregular component of the interplanetary field. According to Jokipii and Davis (1969) the waves may disappear beyond a few AU. However it is by no means clear that the observations require that there should be a boundary to produce an exponential decay. At the higher energies Krimigis (1965) has shown that the model with $\kappa = \kappa_0 r^\beta$ can provide a good fit to the observations of the mean intensity, and he notes that the form (6) approximates exponential behavior for a wide range of t in the decay phase. Burlaga (1967) has shown that for lower energies this model will not fit the observations at very late times as indicated in Figure 3. It is evident however that for this example the model does in fact fit the observations quite well for ~ 60 hr following the flare. It cannot be argued that the lack of agreement at larger times is evidence for the existence of a boundary since the basic assumptions of the isotropic diffusion theory are in any case violated at such times. In the first place one cannot assume that $2\frac{1}{2}$ days after the flare has occurred, the interplanetary medium can be approximated by its pre-flare condition. Secondly, the effects of convection and energy changes become important after about 20–30 hr (depending on the spectral index) so that equations (1) and (2) are not valid (cf. Forman, 1970). It is also possible that an exponential decay could be due to source effects (Reid, 1964; Forman, 1971), anisotropic diffusion (Feit, 1969), or solar rotation (e.g. O'Gallagher, 1970).

Diffusion theories have a defect which can lead to some confusion, namely they assume that the diffusing particles have no inertia. This leads to anomalous predictions such as:

(a) the particles begin to appear at the Earth immediately following the flare, as if they had infinite speed (see Equation (6) for example);

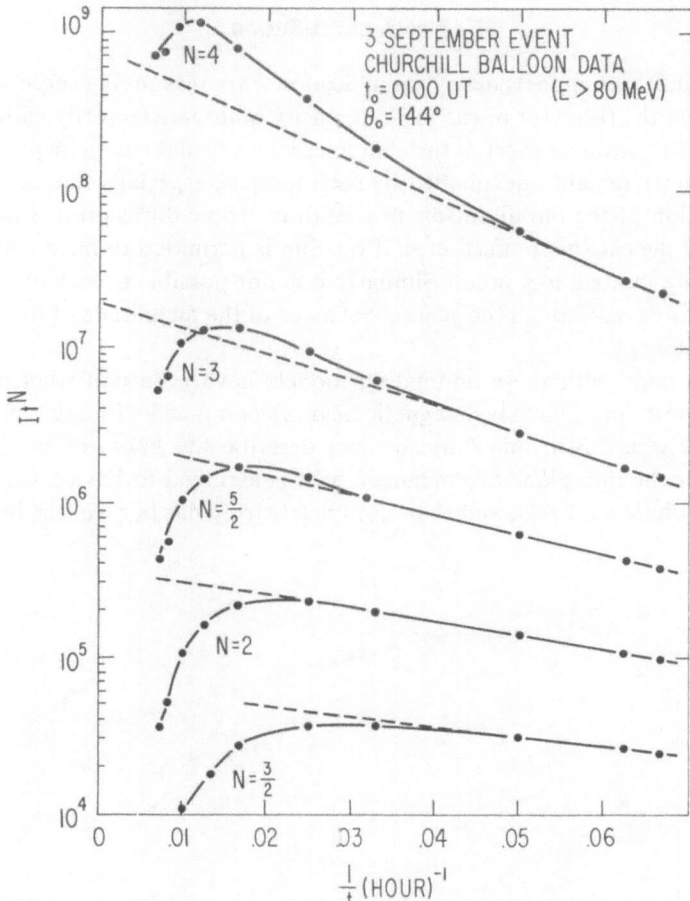

Fig. 3.   A plot of It^N versus $1/t$ for the event of 3 September 1960, with I proportional to the mean intensity j_0, and $N = 3/2(2-\beta)$ (from Burlaga, 1967). It is evident that it is not possible to find a value of β which can organize the data for this event into a straight line in this type of plot, although with $N = \frac{5}{2}$ the fit is not bad up to about $t = 60$ hr. An exponential decay on the other hand, organizes the data very well (see Figure 7).

(b) the anisotropy is initially infinite (Equation (8)) instead of being limited to 100%, which is the maximum physically possible. A more accurate theory based on the telegraph equation rather than the diffusion equation can eliminate both these ano-malies (Axford, 1965b; see especially Equations (7)–(9) of Fisk and Axford, 1969a). Unfortunately the equations are of higher order, since the inertia term $m\partial S/\partial t$ is retained on the left of (2), and thus they are more difficult to solve in general. It appears however that the diffusion equation can be expected to give good results after an elapsed time $t \gg \tau$ and $t \gg r/v$, when the anisotropy falls to less than about 30% (see Figure 4). These effects have also been considered by Shishov (1966), Fibich and Abraham (1965), and Burlaga (1969, 1970).

3. Anisotropic Diffusion

The isotropic diffusion models described in Section 2 are reasonably successful in that they can predict the behavior of the mean intensity quite satisfactorily with a suitable choice of $\kappa(r, T)$. However most of the characteristics of solar energetic particle events listed in Section 1 remain unexplained by such models, especially the east-west effect and the direction of the initial anisotropy. With isotropic diffusion it is not possible to account for the east-west effect, even if the Sun is permitted to have a finite radius instead of being treated as a point. Similarly it is not possible to account for a non-radial anisotropy even though the general behavior of the magnitude of the anisotropy might be correctly predicted.

In order to cope with these anomalies, models have been constructed in which regions of smooth interplanetary magnetic field are combined with regions of tangled magnetic field where isotropic diffusion can describe the behavior of the particle distribution. If the interplanetary magnetic field is assumed to have a smooth spiral form in $r < a$, where $a > 1$ AU, and to be completely irregular in $r > a$, the initial aniso-

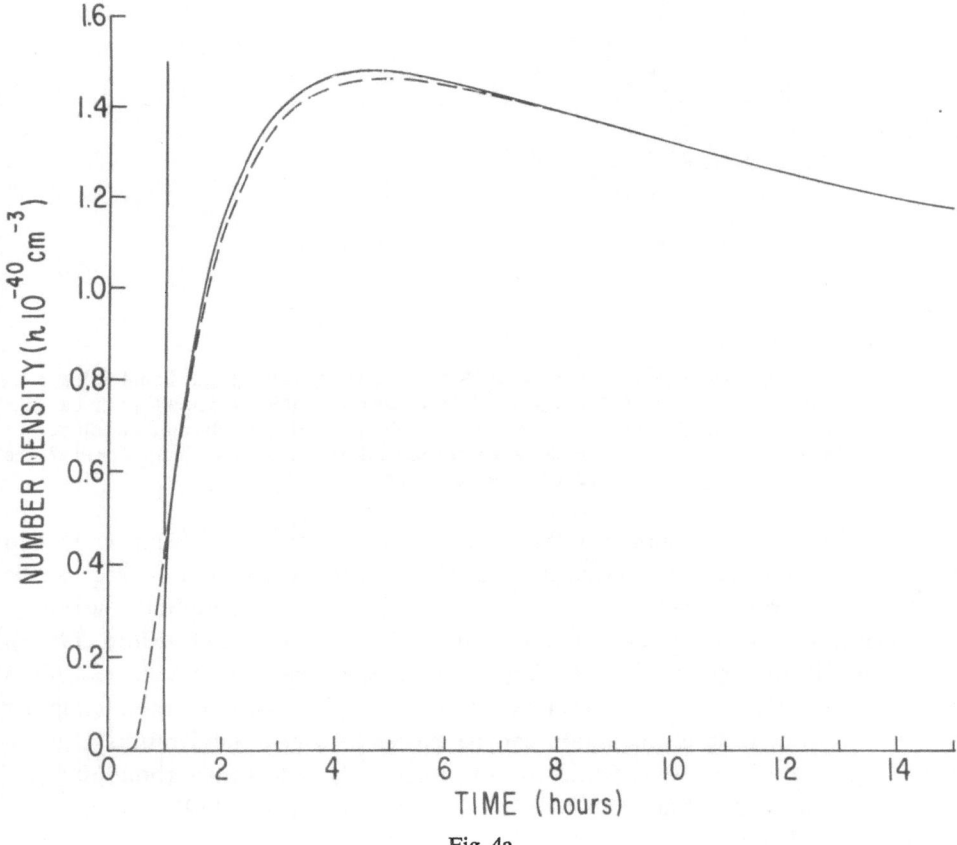

Fig. 4a.

tropy would be aligned parallel to the field as required, provided the source is close to the field line connecting the Sun to the Earth (see Figure 5a). If the source is not so placed, it seems that this model would predict that the initial anisotropy is directed towards the Sun, since the particles would find their way to the Earth by first diffusing through the region $r > a$. By including a free escape boundary at $r = b > a$, with iso- tropic diffusion in $a < r < b$, one can arrange that the decay of the intensity at the Earth is exponential; models of this type have been analyzed extensively by Parker (1963).

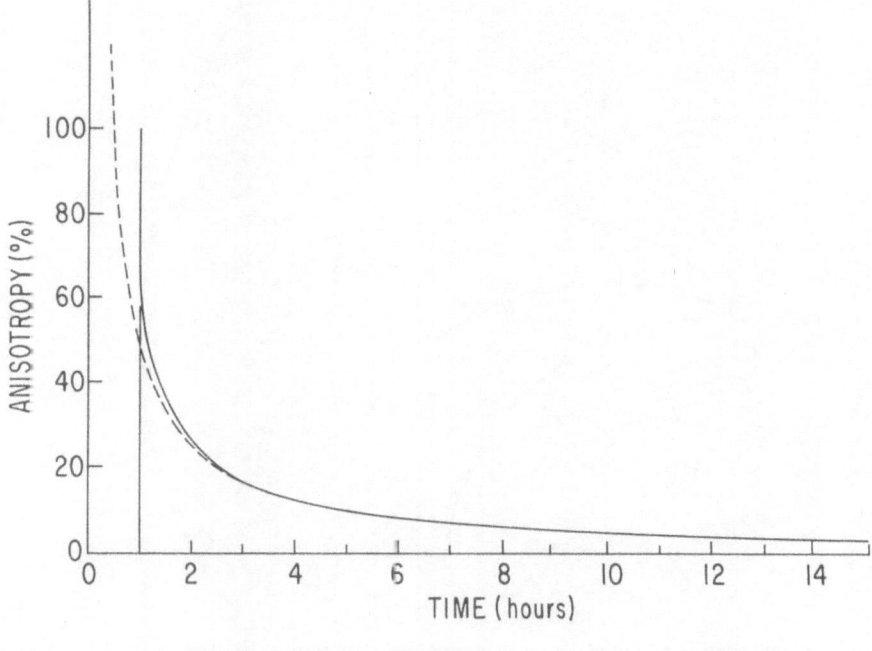

Fig. 4b.

Figs. 4a–b.   A comparison of the predictions of solutions of the telegraph equation (solid lines) and of the diffusion equation (dashed lines) for an idealized model of an energetic solar particle event (see Equations (7)–(12) of Fisk and Axford, 1969a). The behavior of the number density is given in (a) and that of the anisotropy in (b). It is assumed that the energy of the particles is 10 MeV, $\tau v = 0.1$ AU and $r = 1$ AU. Note that the two equations predict essentially the same behavior once the parti- cles have begun to arrive according to the telegraph equation, about 2 hr for the example chosen. The diffusion equation predicts that particles begin to arrive immediately and that the initial aniso- tropy is indefinitely large, both of which effects are physically impossible. The telegraph equation predicts that particles arrive first as a decaying pulse which propagates at the particle speed, and the initial anisotropy is 100%.

To produce the east-west effect, Reid (1964) has postulated that there should be a thin isotropically-diffusing shell surrounding the Sun (see Figure 5b), from which particles are able to escape along smooth interplanetary field lines and so reach the Earth. If the particles escape from the shell at a rate $k U_s$, where $k = k(T)$ is independent of position on the Sun, and U_s is the surface density of particles, then the diffusion equation appropriate to the shell is

$$\frac{\partial U_s}{\partial t} = \frac{\kappa_s}{\varrho} \frac{\partial}{\partial \varrho} \left(\varrho \frac{\partial U_s}{\partial \varrho} \right) - k U_s, \tag{12}$$

where κ_s is the diffusion coefficient, and ϱ is the distance of a point on the shell from the location of the flare. Effects associated with the spherical geometry have been neglected in this formulation, which requires that ϱ should be less than the radius of

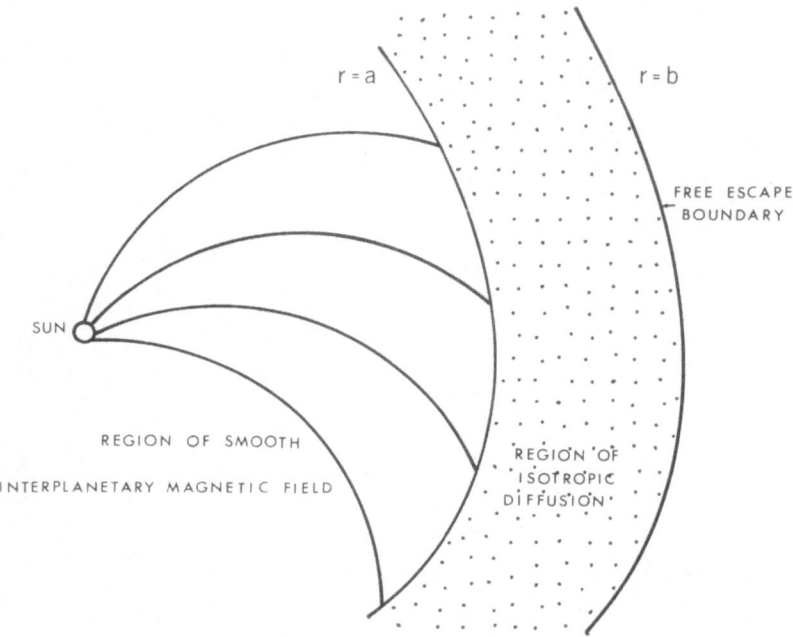

Fig. 5a. Configuration of a model with a smooth interplanetary magnetic field region and an isotropically diffusing shell with a free escape outer boundary (Parker, 1963).

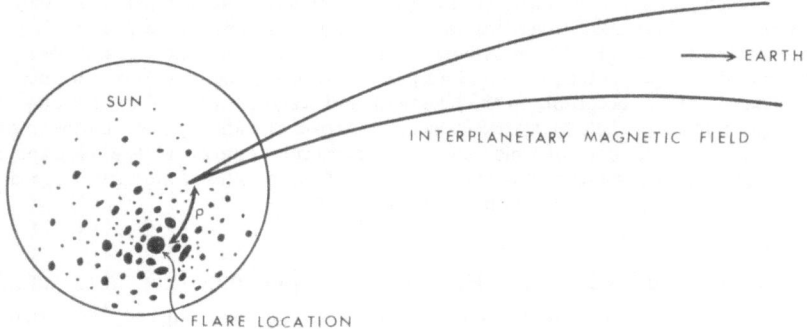

Fig. 5b. Sketch of the physical configuration of Reid's diffusing layer (see also Figures 6 and 8); the dots are intended to represent the particles diffusing away from the flare region (Reid, 1964).

the Sun; this is obviously not true for east limb flares, but the error introduced is not large in view of the nature of the model and the simplification achieved with this assumption. Note that the effects of energy loss could be included in (12) without much difficulty, and that absorption of particles in the chromosphere can be taken into account by increasing k by an appropriate factor.

The solution of (12) for the case in which N particles are released impulsively at $\varrho = 0$, $t = 0$ is

$$U_s(\varrho, t) = \frac{N}{4\pi\kappa_s t} \exp\left\{-\frac{\varrho^2}{4\kappa_s t} - kt\right\}.$$ (13)

The east-west effect is produced in this model as a result of the factor $\exp(-\varrho^2/4\kappa_s t)$, since the effective source for an observer on an interplanetary magnetic field line which has its 'foot' in the lower corona at a distance ϱ from the flare location is $kU_s(\varrho, t)$. If there were no scattering due to magnetic irregularities in the interplanetary region, the intensity observed at the Earth would be proportional to U_s; however the anisotropy would always be 100% which is contrary to what is observed. In order to provide a more satisfactory model, it is necessary to combine Reid's diffusing shell with a model describing the behavior of particles in the interplanetary region. For example a combination of the Reid and Parker models shown in Figure 5 could probably explain all the characteristics of solar energetic particle events listed in Section 1 other than the 'equilibrium' anisotropy, although in addition to N and ϱ, there would be at least five independent parameters available for fitting the observations. One should regard Reid's diffusing layer as a first attempt to account for the storage and diffusion of particles near the Sun, which is believed to occur (e.g. Elliot, 1972); it is possible to construct other models to describe these effects, but they must always contain at least two parameters.

It is also possible to combine Reid's model with a 'bottle' or 'tongue' configuration of the interplanetary magnetic field (Cocconi et al., 1958; Gold, 1959, 1962; Piddington, 1958). In this case 'weak' scattering in the interplanetary region is required to produce the observed decay of the anisotropy, and the intensity is assumed to decay eventually as a result of leakage of particles from the 'bottle' and adiabatic deceleration associated with the expansion of the 'bottle'. Although this type of model has several attractive features, it has not been developed formally and hence it is not possible to make a direct comparison with detailed observations of solar energetic particle events. 'Bottles' must occur occasionally (e.g. Schatten, 1970), but on the other hand there is no evidence that the interplanetary field near the Earth closes back onto the Sun in such a manner that it forms a reasonably compact 'bottle' most of the time.

It is clear that in order to provide good models, anisotropic diffusion must be taken into account. In a first approach to this problem (Axford, 1965a) it was shown that the ratio of the diffusion coefficients perpendicular and parallel to the mean magnetic field should be $\kappa_\perp/\kappa_\parallel \approx 1/(1 + (\omega_c\tau)^2)$, where ω_c is the particle gyro-frequency in the mean field and τ is the relaxation time as before; it is expected that $\omega_c\tau \gg 1$ in general and hence $\kappa_\perp \ll \kappa_\parallel$. A simple anisotropic diffusion model can be constructed assuming

$\omega_c \tau = \infty$ (i.e. $\kappa_\perp = 0$), so that the particles essentially travel along the interplanetary magnetic field lines being scattered in pitch angle. One can choose any form for $\kappa_\|$ provided distance is measured along the field lines (x) instead of radially from the Sun (r). It is necessary to provide some form of transverse diffusion near the Sun so that flares at any longitude can inject energetic particles onto the interplanetary field lines leading to the Earth (see Figure 6). Reid's diffusing layer is especially suitable for

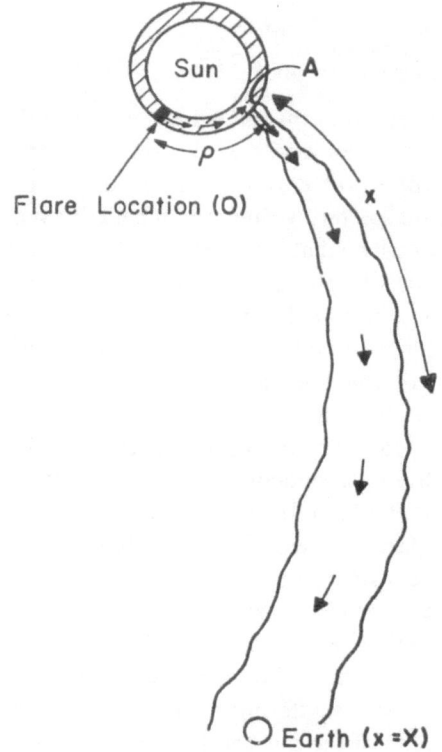

Fig. 6.   A simple model for energetic solar particle events allowing for a diffusing layer near the Sun (Reid, 1964), and escape of particles on to spiral interplanetary magnetic field lines. In the interplanetary region, diffusion is anisotropic with $\kappa_\perp = 0$ (from Axford, 1965b). The arrows indicate the path of particles which eventually reach the Earth.

this purpose, and when combined with an interplanetary diffusion coefficient of the form $\kappa_\| (x) = \kappa_0 x^\beta$, one obtains for the mean intensity near the Earth:

$$j_0 (x_e, t) = A \int\limits_0^t \frac{\exp\{-\alpha/\xi - k\xi - \gamma/(t-\xi)\}}{\xi(t-\xi)^n} \, d\xi, \tag{14}$$

where A is a constant, $n = 3/(2-\beta)$, $\alpha = \varrho^2/4\kappa_s$, $\gamma = x_e^{2-\beta}/(2-\beta)^2\kappa_0$ and $x_e \simeq 1.3$ AU is the distance to the Earth (Axford, 1965b).

The integral in (14) is not very convenient for direct comparison with observations; however, asymptotic approximations suitable for small and large values of the time are relatively simple:

$$j_0(x_e, t) \propto t^{-(n-1/2)} \exp\{-(\alpha^{1/2} + \gamma^{1/2})^2/t\}, \quad t \to 0, \tag{15}$$

$$j_0(x_e, t) \propto t^{-n} \exp\{-\gamma/t\}, \quad t \to \infty. \tag{16}$$

The result for $t \to 0$ given by Axford (1965b) is correct only for $\beta = 0$ (Kinsey, private communication). If the data for early and late times in an event are plotted in a manner similar to that shown in Figure 1, it is possible to determine three of the independent parameters (α, β, γ). The scale factor A is of course easily obtained, while the fourth independent parameter (k) can be determined by trial and error on fitting the complete form (14) to the data for the whole of the event. It should be noted that (16) exhibits the same time-dependence as (6), and hence the determinations of β and γ made by Krimigis (1965) can be used in this anisotropic diffusion theory provided the parameters are suitably interpreted. The anisotropy early and late in the event, is similar to that obtained in the isotropic diffusion model (8), namely

$$\xi = 3[1 + \sqrt{(\alpha/\gamma)}] x/(2 - \beta) vt, \quad t \to 0, \tag{17}$$

$$\xi = 3x/(2 - \beta) vt, \quad t \to \infty; \tag{18}$$

in this case, however, the direction of the anisotropy is always parallel to the interplanetary magnetic field lines and not radial. For eastern hemisphere flares ($\alpha/\gamma = O(1)$), the anisotropy is larger at a given time earlier in the event than it would be for a western hemisphere flare ($\alpha/\gamma \approx 0$). However the intensity increases to maximum more slowly for eastern hemisphere flares (as shown by (15)) and hence the period of large anisotropy is much less noticeable than it is for flares in the western hemisphere. It is evident then, that a model of this type can account in a rough fashion for all the characteristics of solar energetic particle events except the 'equilibrium' anisotropy, which in any case occurs late in the event when the model is no longer valid as a result of the neglect of convection and energy changes. Note that it is possible to make different choices for the form of $\kappa_{\parallel}(x)$, and to allow for a free-escape boundary at $x = a > x_e$; however, it is doubtful whether anything is to be gained by such an exercise, since the goodness of fit between model and observations does not seem to be very sensitive to the nature of the free parameters involved, provided there are enough of them.

A different type of anisotropic diffusion model has been considered by Burlaga (1967), omitting the diffusing layer near the Sun (or its equivalent), but allowing for diffusion in the direction transverse to the mean magnetic field, which is assumed to be radial. The diffusion coefficients are taken to be $\kappa_{\parallel} = $ constant, $\kappa_{\perp} \propto r^2$, together with a free escape boundary at $r = a > 1$ AU. In this case an analytic solution can be obtained in the form of an infinite series which has convenient asymptotic forms for small and large values of the time. By fitting observed data to these asymptotic forms it is possible to determine the scale factor and the other two free parameters involved in the model. An example of such a fit is shown in Figure 7 for the event of 3 September 1960 (also shown in Figure 3). The exponential decay of the mean intensity, associated with the presence of a free escape boundary in this model, will often allow a good fit to the data over a long period of the decay phase. However, as pointed out in Section 2, this does not necessarily have much meaning since the neglect of convection and energy changes renders the model invalid after about 20 hr (see Section 4).

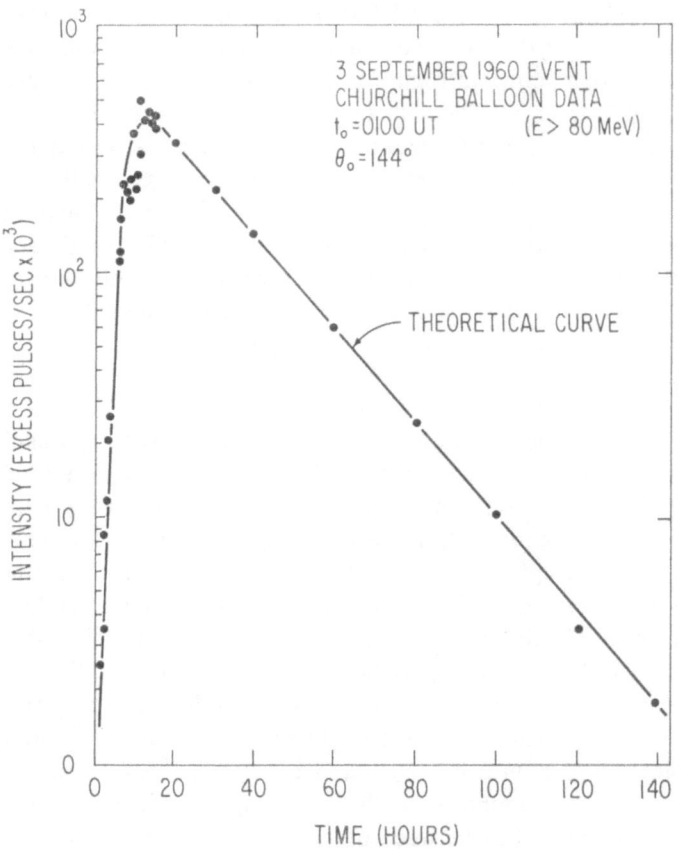

Fig. 7.   Theoretical fit of balloon data for the energetic solar particle event of 3 September, 1960, to the anisotropic diffusion model of Burlaga (1967). The period of exponential decay (20–140 hr) is produced in this model by a free escape boundary situated beyond the orbit of Earth.

The most interesting aspect of Burlaga's model is the inclusion of transverse diffusion. If $\kappa_\perp/\kappa_\parallel = 1/(1+(\omega_c\tau)^2)$ as suggested by the simplest theory of anisotropic diffusion (Axford, 1965a, b), then the choice $\kappa_\parallel = $ constant (i.e. $\tau = $ constant) and $\kappa_\perp/\kappa_\parallel \propto r^2$ would require that $\omega_c \propto 1/r$ which is not very reasonable in $r < 1$ AU. However, Jokipii and Parker (1969) have pointed out that random walk of magnetic field lines determined from the observed power spectrum of interplanetary magnetic field fluctuations extrapolated to zero frequency (Jokipii, 1966), and from the observed turbulent velocity field in the solar photosphere, yields an additional component to $\kappa_\perp$ which is not small if $\omega_c\tau \gg 1$, and which should vary as r^2. Thus we may interpret the transverse diffusion in Burlaga's model as being due to field line random walk, and there is no contradiction between the assumptions $\tau = \text{constant}$ and $\kappa_\perp/\kappa_\parallel \propto r^2$. Other models which can be interpreted in this manner have been analyzed by Feit (1969). It is interesting to note that Feit's solution for the case $\kappa_\parallel \propto r$, $\kappa_\perp \propto r^2$ appears to yield an exponential decay of the mean intensity at large times.

Burlaga finds that typically $\kappa_\perp = 8 \times 10^{21} (r/r_e)^2$ cm^2 sec^{-1} for $T \gtrsim 100$ MeV, which is somewhat larger than the estimate $\kappa_\perp \simeq 2 \times 10^{21}$ v/c cm^2 sec^{-1} at $r = r_e$, obtained from measurements of the power spectrum of interplanetary magnetic field fluctuations (Jokipii and Coleman, 1968; Jokipii and Parker, 1969). Klimas and Sandri (1970) have given a more detailed theory for diffusion coefficients at low energies than previously available, and suggest that $\kappa_\perp$ is probably even smaller than the above estimate. Possibly a more direct method of estimating the amount of field line random walk is to note that the temporal structure of long-lasting events commonly exhibits a 'core' of electrons ($T > 40$ keV) which appears to be associated with particles coming almost directly from the source (Figure 8). The width of this feature is typically only a few degrees of solar longitude, and it presumably provides a measure of the amount of field line random walk between the Sun and the Earth. This suggests that we cannot rely on field line random walk alone to produce the observed spread in longitude of particles emitted from flares and active regions, and hence that something like Reid's diffusing layer must also be considered. Finally, it is not clear that the initial anisotropy would be parallel to the interplanetary magnetic field for events associated with flares in the eastern hemisphere (McCracken et al., 1967) in models such as Burlaga's which rely entirely on transverse diffusion in the interplanetary region to produce longitudinal spreading and the east-west effect. If a diffusing layer near the Sun (or its equivalent) does exist, then it controls the gradients of the mean intensity perpendicular to the mean interplanetary magnetic field direction and accordingly reduces the importance of transverse diffusion in determining the behavior of solar energetic particles (cf. Feit, 1969).

It has been pointed out by Lin et al. (1968) and by O'Gallagher (1970) that for events which last for several days, solar rotation plays an important role in producing the time variations of the flux observed at the Earth (see Figure 8). This effect can be taken into account quite easily in anisotropic diffusion theories by noting that the solar longitude of the Earth with respect to the source of the particles is a linear function of time (e.g. ϱ is a function of t in (14)). From simultaneous observations of low energy (~ 15 MeV) protons from two widely separated spacecraft, O'Gallagher notes that an upper limit can be deduced for the diffusion coefficient perpendicular to the magnetic field lines, viz. $\kappa_\perp \lesssim 1.2 \times 10^{19}$ cm^2 sec^{-1} at 1 AU. This is substantially smaller than the value for $\kappa_\perp$ suggested above, and apparently inconsistent with the observed longitudinal spread of the particles, which would have required $\kappa_\perp \simeq 1.4 \times 10^{20}$ cm^2 sec^{-1} if it were due only to diffusion in the interplanetary region. The peculiar

asymmetry of the proton and electron 'cores' shown in Figure 8 have been interpreted by Jokipii (1969) as being due to gradient and curvature drifts of protons in the
interplanetary magnetic field, and in turn as evidence that the diffusion coefficient
parallel to the interplanetary field must be quite small for the particles concerned
($\kappa_{\parallel} \approx 10^{20}$ cm^2 sec^{-1} for $1-10$ MeV protons).

4. Effects of Energy Changes and Convection

It has been shown by Gleeson and Axford (1967) that if the effects of convection
by the solar wind and the resultant energy changes are taken into account, then

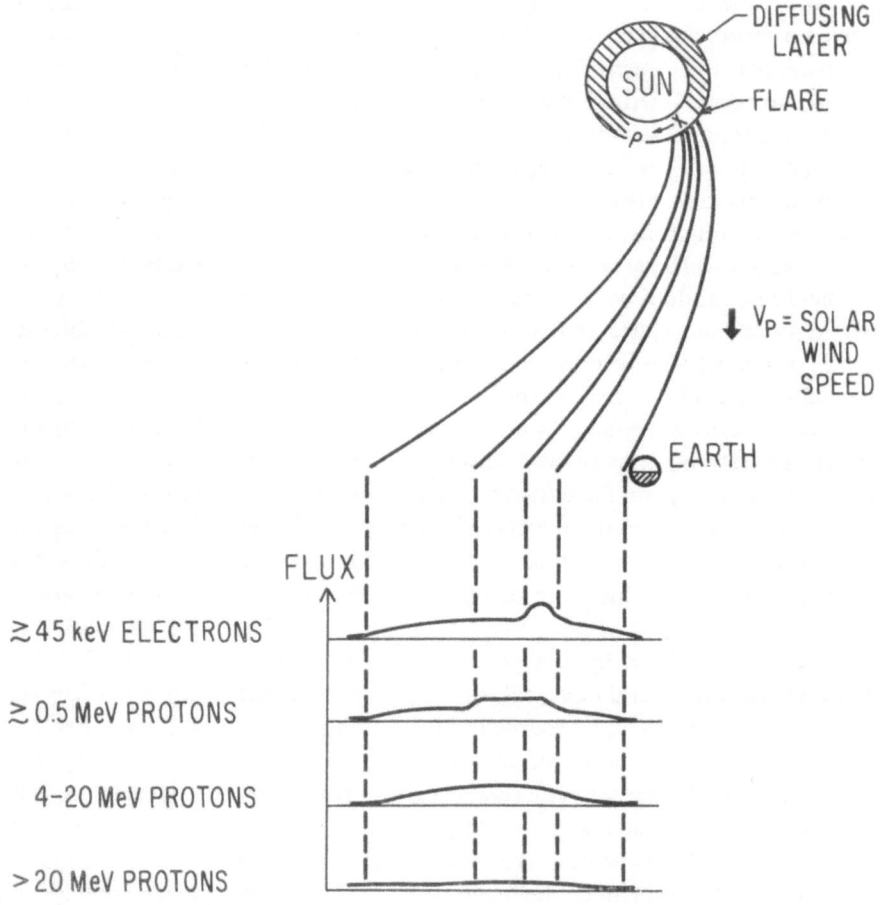

Fig. 8.   A schematic diagram of the core and halo of the energetic particle fluxes in the event of
7 July, 1966. The core and halo are convected by the solar wind past the Earth. The spatial variations
of the particle fluxes are indicated in the lower part of the diagram. Note that the low energy proton
core is not symmetrical about the electron core. The latter is presumed to be connected directly to the
flare region. (Taken from Lin *et al.*, 1968.)

for a spherically-symmetric model Equations (1) and (2) should be replaced by

$$\frac{\partial U}{\partial t} + \frac{1}{r^2}\frac{\partial}{\partial r}(r^2 S) = -\frac{1}{3}V\frac{\partial^2}{\partial r \partial T}(\alpha TU), \tag{19}$$

$$S = CVU - \kappa\frac{\partial U}{\partial r}, \tag{20}$$

where

$$C(r, T) = 1 - \frac{1}{3U}\frac{\partial}{\partial T}(\alpha TU), \tag{21}$$

and $\alpha = (T + 2mc^2)/(T + mc^2)$, with m being the particle rest mass. On eliminating S from these equations, a Fokker-Planck equation given originally by Parker (1965) is obtained for U;

$$\frac{\partial U}{\partial t} + \frac{1}{r^2}\frac{\partial}{\partial r}(r^2 VU) - \frac{1}{3r^2}\frac{\partial}{\partial r}(r^2 V)\frac{\partial}{\partial T}(\alpha TU) = \frac{1}{r^2}\frac{\partial}{\partial r}\left(r^2\kappa\frac{\partial U}{\partial r}\right) \tag{22}$$

which differs from (4) by the appearance of the terms involving V. A generalization of this equation allowing for anisotropic diffusion has been given by Jokipii and Parker (1970).

If we assume α to be constant and $U \propto T^{-\mu}$, then if κ is independent of T the energy spectrum is preserved at all times, and $C = \{1 + \frac{1}{3}\alpha(\mu - 1)\}$. For non-relativistic particles $\alpha \approx 2$ and $\mu = \gamma + \frac{1}{2}$, where γ is the spectral index corresponding to the mean intensity j_0, and hence for $\gamma \approx 3-6$ we find that $C \simeq 2.3-4.3$. Thus the effects are more substantial than one would expect from convection with no energy changes, when $C = 1$ (e.g. Budilov et al., 1963). The significance of the term CVU in (20) can be estimated by comparing it with $\kappa\,\partial U/\partial r$ for a model such as that described by Equation (6); in this case

$$D = \left|CVU/\kappa\frac{\partial U}{\partial r}\right| = (2 - \beta)\,CVt/r, \tag{23}$$

and hence, with $\beta \approx 1$ and $V \approx 400$ km sec^{-1}, we find that $D \approx 1$ when $t \approx 20-45$ hr (depending on γ). Note that for relativistic particles ($\alpha \approx 1$), $C \approx 1.8-2.5$ and $D \approx 1$ when $t \approx 35-60$ hr, so the effects of convection and energy changes are much less important since the events do not usually last for this length of time.

It is clear that the effects of convection and energy changes must be taken into account for non-relativistic particle events which last for times of the order of a day or more. A solution of the above equations which illustrates these effects has been given by Fisk and Axford (1968) for the case $\kappa = \kappa_0 r$, $\alpha = 2$, with initial conditions $U(r, T, t) = AT^{-\mu}\delta(r - r_0)/r^2$ at $t = 0$:

$$U(r, T, t) = \frac{AT^{-\mu}}{\kappa_0 trr_0}\left(\frac{r}{r_0}\right)^{V/2\kappa}\exp\left[-(r + r_0)/\kappa_0 t\right]I_\eta(2\sqrt{rr_0}/\kappa_0 t), \tag{24}$$

where $\eta = [(2 + V/\kappa_0)^2 + 16V(\mu - 1)/3\kappa_0]^{1/2}$, and $I_\eta(\varphi)$ is the modified Bessel function of the first kind and order η. For large times $(t \gg 2(rr_0)^{1/2}/\kappa_0)$, the asymptotic form of (24) is

$$U(r, T, t) \sim \frac{AT^{-\mu}(r/r_0)^{V/2\kappa_0}}{\Gamma(1+\eta)(\kappa_0 t)^{1+\eta}(rr_0)^{1-\eta/2}} \exp(-(r+r_0)/\kappa_0 t). \qquad (25)$$

Thus for large t, $U(r, T, t) \propto t^{-1-\eta}$, which is to be compared with the result $U(r, T, t) \propto t^{-3}$ obtained in the absence of convection and energy changes ($V=0$ in (25) or $\beta=1$ in (6)). Accordingly if $\mu=4$, and $V \simeq 400$ km sec^{-1}, the effects of convection and energy changes become very pronounced indeed when $\kappa < 10^{21}$ cm^2 sec^{-1} at $r=1$ AU.

The corresponding solution for the anisotropy at large times is

$$\xi_r \sim (3V/2v)\left[1 + \tfrac{4}{3}(\mu-1) + (2-\eta)\kappa_0/V + 2r/Vt\right]. \qquad (26)$$

In contrast to the result (8) for the case where convection and energy changes were neglected, there is a residual anisotropy at large times which approaches $2(\mu-1)V/v$ when $V/\kappa_0 \gg 1$, together with a time-dependent term which is the same as (8) with $\beta=1$. The residual anisotropy is $3(C-1)V/v$ rather than $3CV/v$ which would be expected if purely convective transport were to dominate in (20). Thus, in this case at least the effects of diffusion are still important late in the event even when the diffusion coefficient is small. Note that the contribution of diffusion to the anisotropy is negative at large times due to the fact that convection alters the radial distribution in such a manner that $\partial U/\partial r > 0$ (cf. Parker, 1966; Fisk and Axford, 1968). The point of maximum density moves at speed $\tfrac{1}{2}(V+\kappa_0(\eta-2)) \to V$ as $\kappa_0 \to 0$.

This particular model, although informative, does not represent the observations very well, since like the isotropic diffusion models described in Section 2 it cannot explain the east-west effect and the direction of the initial anisotropy. Furthermore the power-law decay predicted for large t is too steep to fit the observations, especially at low energies where V/κ_0 is expected to be large (cf. Krimigis, 1965). At least some of the deficiencies of the model could be remedied, however, by introducing it as part of the anisotropic diffusion model of Axford (1965b), taking $\kappa_{\parallel}=\kappa_0 x$ and $\kappa_{\perp}=0$, and making use of a thin diffusing layer near the Sun to provide the dependence on solar longitude. In this case some care must be taken in distinguishing between r and x and in performing the transformation between the frame fixed in the rotating Sun and the non-rotating frame in determining the various components of the anisotropy.

An alternative approach has been adopted by Forman (1970b), who in effect combines Burlaga's (1967) model for anisotropic diffusion with a free-escape boundary with the above model which allows for convection and energy loss. The diffusion coefficients in Forman's model are $\kappa_r = \kappa_{\parallel} = \kappa_0 r$ and $\kappa_{\perp} = \kappa_1 r^2$, which permit an analytic solution to be obtained by writing

$$U(r, \theta, t, T) = \Theta(\cos\theta, t) R(r, t) AT^{-\mu}, \qquad (27)$$

and using the method of separation of variables. The inclusion of a free-escape boundary at $r=a$ produces an exponential decay with time constant $t_D = a/2CV$, which is independent of T (in contrast with (10) where the time constant depends on T through

κ_0). For a wide range of values of κ_0 (10^{19}–10^{21} cm^2 sec^{-1}) the decay time is of the order of 10–18 hr, if $a = 2.3$ AU. At large times the anisotropy is essentially the same as (26) at points not too close to the boundary (where the anisotropy increases rapidly as in (11)). In this case the point of maximum density moves away from the Sun at a speed of order V initially, but ultimately it must dwell quite close to the boundary. Forman also shows that the decay of the mean intensity with time could be due to source effects if the leakage time from the source $1/k$ (cf. (13)) is such that $kt_D \ll 1$, but notes that it would then be a pure coincidence that the observed decay times are of order $r/2CV$.

An extensive numerical study of this problem has been carried out by Englade (1970) using a model which can be regarded as a combination of all the models we have discussed. Unfortunately, with so many free parameters available it is not easy to determine which ones are the most important. Nevertheless it is clear that a quite reasonable model of a solar energetic particle event can be generated in this way. One interesting feature of the results is that it is confirmed that longitudinal diffusion near the Sun is necessary if the direction of the initial anisotropy of particles emitted by eastern hemisphere flares is to be more or less parallel to the interplanetary magnetic field as observed. The results for the decay phase are consistent with the predictions of Forman's model as would be expected from the configuration chosen.

5. Continuous Emission of Particles from the Sun

There is a growing body of evidence which suggests that the Sun is an important source of particles in the low energy range $T \lesssim 50$ MeV/nucleon (e.g. Vogt, 1962; Kinsey, 1970; Krimigis and Venkatesan, 1969; Krimigis, 1969; Vernov et al., 1969; Fan et al., 1968, 1969). The suggestion is based on the temporal behavior of the cosmic ray spectrum at low energies, and the radial gradients and anisotropies of the particles concerned. It is therefore of interest to consider the relatively simple problem of steady emission of particles by the Sun taking into account the various processes which must be important such as diffusion, convection by the solar wind and the energy losses associated with convection.

First we note that in the absence of scattering, particles released at the Sun would move outwards along interplanetary magnetic field lines conserving their magnetic moment. At the orbit of Earth, the particles would have very small pitch angles, and as a result of co-rotation of the interplanetary magnetic field with the Sun would have slightly increased energies. If the interplanetary magnetic field were purely radial and not co-rotating with the Sun, the anisotropy at the Earth would be 100% and the radial gradient of the particle density would be $(r/U)(\partial U/\partial r) = -200\%$ per AU.

If the interplanetary medium were spherically-symmetric and we take diffusion into account but ignore convection and energy changes, then (1) yields

$$r^2 S = \Phi/4\pi, \tag{28}$$

where $\Phi (T)$ is the total flux emitted by the Sun in the kinetic energy range $(T, T+\mathrm{d}T)$.

The particle density satisfies the equation

$$\kappa \frac{dU}{dr} = -\Phi/4\pi r^2, \tag{29}$$

which can be solved easily for given κ (r, T). In particular if we take $\kappa = \kappa_0 r$ we find that

$$U = \Phi/8\pi\kappa_0 r^2. \tag{30}$$

The anisotropy and radial gradient of the mean density (or intensity) are given by

$$\xi_r = 6\kappa_0/v, \tag{31}$$

$$\frac{r}{U}\frac{dU}{dr} = \frac{r}{j_0}\frac{dj_0}{dr} = -2. \tag{32}$$

Thus the anisotropy is constant (and presumably much less than 100%), and the radial gradient is again -200% per AU at the orbit of Earth. It should be a relatively straightforward matter to generalize this type of model to take into account the effects of anisotropic diffusion in a co-rotating spiral interplanetary magnetic field. If energy changes are neglected, the effects of convection by the solar wind can also be included without much difficulty, especially for the case $\kappa = \kappa_0 r$ (e.g. Parker, 1963).

If we wish to take the effects of energy changes into account, using Equations (19), (20) and (22) with $\partial U/\partial t = 0$, it is less easy to obtain solutions. However, if we assume that α is constant, that $\kappa = \kappa(r)$, and that the spectrum of the particles is a power law, then $U(r, T) = AT^{-\mu}f(r)$ where $f(r)$ satisfies the ordinary differential equation

$$\frac{d}{dr}\left\{r^2\left(\kappa\frac{df}{dr} - Vf\right)\right\} - \frac{d}{dr}(r^2 V)(C-1)f = 0, \tag{33}$$

with $C = 1 + \frac{1}{3}\alpha(\mu - 1)$, and $f(r_0) = 1$. For the case $\kappa = \kappa_0 r$ and V constant, $f(r) = (r/r_0)^{-q}$ where

$$\kappa_0 q^2 - (2\kappa_0 - V)q - 2CV = 0, \tag{34}$$

and hence if $V/\kappa_0 \ll 1$, $q \approx 2$ (in agreement with (30)), and if $V/\kappa_0 \gg 1$, $q = 2C$. These limiting results correspond to situations in which diffusion dominates and the convection and energy change terms in (22) are negligible ($V/\kappa_0 \ll 1$), and conversely ($V/\kappa_0 \gg 1$).

One should expect that in general if $CV\tilde{r}/\kappa \ll 1$ in a region where the characteristic length $\tilde{r} = U/|\partial U/\partial r|$, then the behavior of the particles can be approximated by the simple diffusion Equations (1) and (2) with V placed equal to zero and $\partial U/\partial t = 0$. Similarly if $CV\tilde{r}/\kappa \gg 1$, then (20) can be adequately approximated by

$$S = CVU, \tag{35}$$

and the term involving κ can be neglected in (22), which becomes

$$\frac{\partial}{\partial r}(r^2 VU) = \frac{1}{3}\frac{\partial}{\partial r}(r^2 V)\frac{\partial}{\partial T}(\alpha TU). \tag{36}$$

In the latter case the radial variation of U is entirely determined by convection and energy changes, and if V and α can be assumed constant, (36) can be integrated to yield

$$U(r, T) = \left(\frac{r}{r_1}\right)^{(2\alpha/3)-2} U\left[r_1, \left(\frac{r}{r_1}\right)^{(2\alpha/3)} T\right], \tag{37}$$

where the spectrum is assumed to be known at some point $r=r_1$, (Fisk and Axford, 1969b; Gleeson, 1970). The anisotropy and radial gradient are respectively given by

$$\xi_r = 3CV/v, \tag{38}$$

and

$$\frac{r}{U}\frac{\partial U}{\partial r} = \frac{r}{j_0}\frac{\partial j_0}{\partial r} = -2C, \tag{39}$$

and thus both these quantities can be determined immediately if the spectrum is known. It is important to note that since diffusion does not play a role in this limiting situation, it is of no significance that there is a magnetic field and that the diffusion coefficient is likely to be anisotropic; thus the results (37), (38) and (39) are of quite general validity provided only that κ is sufficiently small.

Gleeson *et al.* (1970) have examined observations of protons with $T \geqslant 0.3$ MeV made from spacecraft at heliocentric distances of 0.84 AU (Mariner 5) and 1.01 AU (Explorer 33) on the basis of the simple model outlined above. They find that the anisotropy and radial gradient observed during quiet periods is indeed consistent with estimates based on (38) and (39), and that the evolution of the spectrum with radial distance is satisfactorily predicted by (37) (see Figure 9). The condition $CV\bar{r}/\kappa \gg 1$ requires that $\kappa \ll 4 \times 10^{20}$ cm^2 sec^{-1}, which is probably reasonable for these low energy particles.

It should be noted that this approximate theory is not expected to be valid everywhere, and in particular one cannot extrapolate the spectrum all the way back to the Sun using (37). In fact it is probable that diffusive propagation is more important near the Sun and that purely convective transport is dominant only beyond a certain heliocentric distance which could be a substantial fraction of 1 AU. This behavior is simply illustrated by the case with κ and V both constant in (33) (Fisk, private communication);

$$U(r, T) = \frac{AT^{-\mu}\mathcal{U}(2C; 2; Vr/\kappa)}{\mathcal{U}(2C; 2; Vr_0/\kappa)}, \tag{40}$$

where $\mathcal{U}$ is a confluent hypergeometric function (Slater, 1960). The asymptotic forms of this result for large and small values of Vr/κ are

$$U(r, T) \sim AT^{-\mu}(r/r_0)^{-1} \ldots, \quad Vr/\kappa \ll 1, \tag{41}$$

$$U(r, T) \propto AT^{-\mu}r^{-2C}\{1 - 2C(2C-1)\kappa/Vr \ldots\}, \quad Vr/\kappa \gg 1. \tag{42}$$

As expected (41) is the solution of (29) with a suitable definition for Φ, and the first term in (42) is the solution of (39). Note that the convergence of (42) is in fact rather slow and that the condition for neglect of the higher order terms is $Vr/\kappa \gg 2C(2C-1)$.

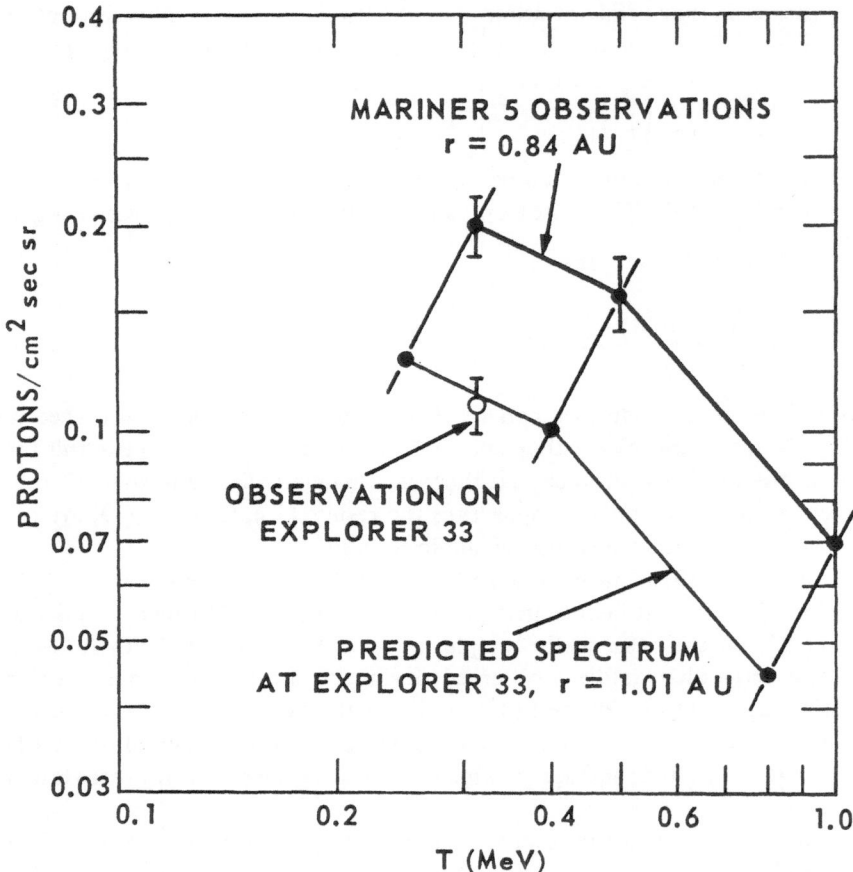

Fig. 9.   The integral intensity spectrum at the position of Explorer 33 ($r = 1.01$ AU) predicted from that observed on Mariner 5 at $r = 0.84$ AU on August 27–31, 1967. The integral intensity actually observed on Explorer 33 for $T \geqslant 0.31$ MeV is shown for comparison. (Taken from Gleeson *et al.*, 1970)

At large heliocentric distances we expect κ to increase rapidly, and hence it is likely that Vr/κ becomes small once more, so that convective transport is limited to a shell-like region (Gleeson, 1970).

6. Conclusions

On the whole it seems reasonable to assert that satisfactory progress has been made in producing models for the propagation of energetic solar particles. We have a good understanding of many of the important physical processes involved, and there is probably not much to be gained by constructing more complicated ad hoc models of the interplanetary medium and the solar environment with ever-increasing numbers of free parameters to allow a match to the observations.

The essential problem is simply that we do not know the facts about the variation of the components of the diffusion tensor with position in the interplanetary medium, and we need observations rather than guess work. Unfortunately, observations made near the orbit of Earth are very difficult to interpret since they involve the integrated effects of many processes occurring in the interplanetary medium and near the Sun. Thus it would be useful to make observations closer to the Sun (e.g. in the vicinity of the orbit of Mercury), and at greater distances (beyond the orbit of Mars). It is a great advantage if particles other than protons are also observed, since these might allow us to distinguish between source and propagation effects.

From the theoretical point of view more attention should be paid to the problems associated with storage and diffusion near the Sun, particularly with regard to energy losses and changes in composition (cf. Englade, 1970; Krimigis and Verzariu, 1970). The nature of 'spikes' and energetic storm particle events has been largely ignored until recently (e.g. Fisk, 1970; Arens, 1970), and there is scope for further work. Finally it should be noted that there are some phenomena which are not satisfactorily explained at the present time, notably the occasional occurrence of persistent, large ($\sim 100\%$) anisotropies (see however, Jokipii, 1968b), and of 'bidirectional' anisotropies (e.g. McCracken et al., 1967).

References

Anderson, K. A.: 1969, rapporteur paper, Proc. 11th. Int. Conf. Cosmic Rays, Budapest.
Arens, J. F.: 1970, submitted to J. Geophys. Res.
Armstrong, T. P., Krimigis, S. M., and Behannon, K. W.: 1970, J. Geophys. Res. 75, 5980.
Axford, W. I.: 1965a, Planetary Space Sci. 13, 115.
Axford, W. I.: 1965b, Planetary Space Sci. 13, 1301.
Axford, W. I. and Reid, G. C.: 1962, J. Geophys. Res. 67, 1692.
Biswas, S. and Fichtel, C. E.: 1965, Space Sci. Rev. 4, 709.
Bryant, D. A., Cline, T. L., Desai, U. D., and McDonald, F. B.: 1962, J. Geophys. Res. 67, 4983.
Bryant, D. A., Cline, T. L., Desai, U. D., and McDonald, F. B.: 1965, Astrophys. J. 141, 478.
Budilov, V. K., Dorman, L. I., Ivanov, V. I., Kolomeets, E. V., and Miroshnichenko, L. I.: 1963, Proc. 8th. Int. Conf. Cosmic Rays, Jaipur.
Burlaga, L. F.: 1967, J. Geophys. Res. 72, 4449.
Burlaga, L. F.: 1969, Proc. 11th. Int. Conf. Cosmic Rays, Budapest.
Burlaga, L. F.: 1970, Solar Phys. 12, 317.
Cocconi, G., Gold, T., Greisen, K., Hayakawa, S., and Morrison, P.: 1958, Nuovo Cimento Suppl. Ser. X, 8, 161.
Desai, U. D. and McDonald, F. B.: 1970, submitted to J. Geophys. Res.
Dolginov,, A. Z. and Toptygin, I. N.: 1968, Icarus 8, 54.
Dorman L. I.: 1963, Progress in Elementary Particle and Cosmic Ray Physics, 7.
Elliot, H.: 1972, in E. R. Dyer (ed.), 'The Sun', Solar Terrestrial Physics/1970, Part I, p. 134.
Englade, R. C.: 1970, submitted to J. Geophys Res.
Fan, C. Y., Gloeckler, G., McKibben, B., Pyle, K. R., and Simpson, J. A.: 1968, Can. J. Phys. 46, S498.
Fan, C. Y., Gloeckler, G., McKibben, B., and Simpson, J. A.: 1969, Proc. 11th. Int. Conf. Cosmic Rays, Budapest.
Feit, J.: 1969, J. Geophys. Res. 74, 5579.
Fibich, M. and Abraham, P. B.: 1965, J. Geophys. Res. 70, 2475.
Fichtel, C. E. and McDonald, F. B.: 1967, Ann. Rev. Astron. Astrophys. 5, 351.
Fisk, L. A.: 1970, submitted to J. Geophys. Res.
Fisk, L. A. and Axford, W. I.: 1968, J. Geophys. Res. 73, 4396.
Fisk, L. A. and Axford, W. I.: 1969a, Solar Phys. 7, 486.

Fisk, L. A. and Axford, W. I.: 1969b, *J. Geophys. Res.* **74**, 4973.
Forman, M. A.: 1970, *J. Geophys. Res.* **75**, 3147.
Forman, M. A.: 1971, *J. Geophys. Res.* **76**, 759.
Gleeson, L. J.: 1970, in preparation.
Gleeson, L. J. and Axford, W. I.: 1967, *Astrophys. J.* **149**, L 115.
Gleeson, L. J., Krimigis, S. M., and Axford, W. I.: 1970, submitted to *J. Geophys. Res.*
Gold, T.: 1959, *J. Geophys. Res.* **64**, 1665.
Gold, T.: 1962, *J. Phys. Soc. Japan* **17**, Suppl. A2, 600.
Hall, D. E. and Sturrock, P. A.: 1967, *Phys. Fluids* **10**, 2620.
Hasselmann, K. and Wibberenz, G.: 1968, *Z. Geophys.* **34**, 353.
Hofmann, D. J. and Winckler, J. R.: 1963, *J. Geophys. Res.* **68**, 2067.
Jokipii, J. R.: 1966, *Astrophys. J.* **146**, 480.
Jokipii, J. R.: 1967, *Astrophys. J.* **149**, 405.
Jokipii, J. R.: 1968a, *Astrophys. J.* **152**, 671.
Jokipii, J. R.: 1968b, *Astrophys. J.* **152**, 997.
Jokipii, J. R.: 1969, Proc. 11th. Int. Conf. Cosmic Rays, Budapest.
Jokipii, J. R. and Coleman, P. J.: 1968, *J. Geophys. Res.* **73**, 5495.
Jokipii, J. R. and Davis, L.: 1969, *Astrophys. J.* **156**. 1101.
Jokipii, J. R. and Parker, E. N.: 1969, *Astrophys. J.* **155**, 777.
Jokipii, J. R. and Parker, E. N.: 1970, *Astrophys. J.* **160**, 735.
Kinsey, J. H.: 1970, *Phys. Rev. Letters* **24**, 246.
Klimas, A. and Sandri, G.: 1970, submitted to *Phys. Rev. Letters.*
Krimigis, S. M.: 1965, *J. Geophys. Res.* **70**, 2943.
Krimigis, S. M.: 1969, Proc. 11th. Int. Conf. Cosmic Rays, Budapest.
Krimigis, S. M. and Venkatesan, D.: 1969, *J. Geophys. Res.* **74**, 4129.
Krimigis, S. M. and Verzariu, P.: 1971, *J. Geophys. Res.* **76**, 792.
Lin, R. P. Kahler, S. W., and Roelof, E. C.: 1968, *Solar Phys.* **4**, 338.
McCracken, K. G.: 1963, in *Solar Proton Manual*, NASA Tech. Rept. R-169, (ed. by F. B. McDonald)
McCracken, K. G., Rao, U. R., and Bukata, R. P.: 1967, *J. Geophys Res.* **72**, 4293.
McCracken, K. G., Rao, U. R., and Ness, N. F.: 1968, *J. Geophys. Res.* **73**, 4159.
O'Gallagher, J. J.: 1970, *J. Geophys. Res.* **75**, 1163.
Ogilvie, K. W. and Arens, J. F.: 1971, *J. Geophys. Res.* **76**, 13.
Parker, E. N.: 1963, *Interplanetary Dynamical Processes*, Interscience Publishers, New York.
Parker, E. N.: 1965, *Planetary Space Sci.* **13**, 9.
Parker, E. N.: 1966, *Planetary Space Sci.* **14**, 371.
Piddington, J. H.: 1958, *Phys. Rev.* **112**, 589.
Quenby, J. J., Balogh, A., Engel, A. R., Elliot, H., Hedgecock, P. C., Hynds, R. J., and Sear, J. R.: 1969, Proc. 11th. Int. Conf. Cosmic Rays, Budapest.
Rao, U. R., Allum, F. R., Bartley, W. C., Palmeira, R. A. R., Harries, J. A., and McCracken, K. G.: 1969, *Solar Flares and Space Research*, North Holland Publ. Co., Amsterdam, p. 267.
Reid, G. C.: 1964, *J. Geophys. Res.* **69**, 2659.
Roelof, E. C.: 1966, Thesis, University of California, Berkeley.
Roelof, E. C.: 1968, *Can. J. Phys.* **46**, S990.
Schatten, K. H.: 1971, *Solar Phys.*, in press.
Shishov, V. I.: 1966, *Geomagnetizm i Aeronomiya* **4**, 223.
Slater, L. J.: 1960, *Confluent Hypergeometric Functions*, Cambridge University Press, London.
Vernov, S. N., Chudokov, A. E., Vakulov, P. V., Gorchakov, E. V., Kontor, N. N., Logachev, Yu. I., Lyubimov, G. P., Pereslegina, N. V., and Timofeev, G. A.: 1969, Proc. 1th. Int. Conf. Cosmic Rays, Budapest.
Vogt, R.: 1962, *Phys. Rev.* **125**, 366.

DISCONTINUITIES AND SHOCK WAVES IN THE INTERPLANETARY MEDIUM AND THEIR INTERACTION WITH THE MAGNETOSPHERE

LEONARD F. BURLAGA

*Laboratory for Extraterrestrial Physics, NASA-Goddard Space Flight Center,
Greenbelt, Maryland U.S.A.*

Abstract. The discontinuous structure of the solar wind is described with emphasis on properties related to geomagnetic impulses. Some of the discontinuities are clearly hydromagnetic shocks and tangential discontinuities, and can produce a significant change in the momentum flux at the magnetosphere boundary. Such a change generates an impulse which propagates through the magnetosphere to the Earth where it is observed world-wide as an impulse in magnetograms. The propagation process is not reviewed here, but the relation between the initial cause (discontinuity) and the final effect (geomagnetic impulse) is reviewed in detail. The various types of impulses are examined, and are related qualitatively to the various types of discontinuities. The magnitude of an impulse is related to the change in the momentum flux. The propagation time and the rise time depend on the propagation process rather than on the initial state. Double shocks have not been observed, but a reverse shock has been identified. Giant pairs can be caused by a shock followed by a tangential discontinuity, and regular pairs may be due to complementary tangential discontinuities.

1. Introduction

Impulsive changes in the geomagnetic field have been extensively studied for many years. Several types have been identified and much is known about their morphology, but the results are somewhat obscured by the proliferation of different and sometimes conflicting notations. There are also many speculations in the literature, some correct and some incorrect, concerning the causes of the impulses.

The advent of space probes has led to the discovery of several kinds of hydromagnetic discontinuities in the solar wind, some of which were shown to cause geomagnetic impulses. In principle, it is now possible to determine unambiguously the causes of geomagnetic impulses and the effects of interplanetary discontinuities on the Earth's field. Many correlations have already been published.

The aim of this review is to present a synthesis of the published observations which definitively shows the relations between interplanetary discontinuities and geomagnetic impulses. The work necessarily falls short of this goal because the observations are incomplete, but the shortcomings show where effort should be concentrated in future observational studies.

Section 2 presents a summary of work concerning the interplanetary discontinuities with an emphasis on properties relevant to the study of geomagnetic impulses. Section 3 summarizes the types of geomagnetic impulses, and emphasizes the kinds of impulses that unambiguously occur world-wide. Section 4 then reviews the simultaneous observations of geomagnetic and interplanetary discontinuities. The geomagnetic impulse is regarded as a final effect and the interplanetary discontinuity as an initial cause and we aim at showing the relations between them.

Dyer (ed.), Solar Terrestrial Physics/1970: Part II, 135–158. All Rights Reserved.
Copyright © 1972 by D. Reidel Publishing Company, Dordrecht-Holland.

2. Interplanetary Discontinuities

2.1. EXISTENCE OF HYDROMAGNETIC DISCONTINUITIES IN THE SOLAR WIND

Direct measurements of the solar wind show that the magnetic field and plasma parameters may change by more than 50% over a distance of $\approx 10^{-5}$ AU. Such a change is
essentially discontinuous on a scale of 1 AU, or even on a scale of 0.01 AU where it is
seen most clearly (see Figure 1). It is found that at least some of these changes have the
characteristics of hydromagnetic discontinuities. The mere existence of such discontinuities is not surprising, for they were predicted long ago from the equations of magnetohydrodynamics. However, it is of fundamental significance that such discontinuities occur in the interplanetary plasma, which is essentially *collisionless*. This shows
that the theory of magnetohydrodynamics is applicable (at least in some instances) to
the solar wind at 1 AU, and reveals an extension of the fluid concept.

Possible types of hydromagnetic discontinuities in an isotropic plasma are as
follows: (1) tangential discontinuities (T.D.'s); (2) contact surfaces; (3) rotational
discontinuities (R.D.); (4) fast shocks; (5) slow shocks.

These are discussed formally in various textbooks (Landau and Lifshitz, 1960;
Ferraro and Plumpton, 1966), and reviews (Colburn and Sonett, 1966; Spreiter and
Alksne, 1969), so we need not go into the mathematical details. We shall, however,
discuss the qualitative characteristics of the different types of discontinuities, their
significance with respect to geomagnetic impulses, and their existence in the solar
wind.

The concept of a *tangential discontinuity* is illustrated in Figure 2. It is an observable
surface (a current sheet in fact) that separates two physically distinct plasmas. On both
sides there is a magnetic field which is *parallel* to the surface but otherwise arbitrary.
The plasma and magnetic field on each side can have any value, subject to the constraint that the pressure, $\sum_i n_i k T_i + B^2/(8\pi)$ (the sum is over all particle species), is the
same on both sides of the discontinuity. A T.D. does not propagate relative to the
solar wind, i.e., there is no mass flux through the surface. But the material on side 1
can move relative to that on side 2 along the surface (hence the term 'glide plane' for
the surface – Burlaga, 1968). Burlaga (1968) has classified T.D.'s into 13 types, according to the sign of the change in B, proton density (n), and proton temperature (T).
The symbol $(+, -, 0)$ denotes an increase in B, a decrease in n, no change in T;
$(0, +, -)$ means no change in B, an increase in n, a decrease in T; etc. It should be
emphasized that such a signature is not a sufficient condition for identification of a
T.D. A change in momentum flux of the solar wind relative to the Earth, $\Delta(nmV_w) \approx$
$\approx mV_w \Delta n$, in which V_w is the solar wind speed (u in some of the figures) occurs across
T.D.'s with signatures $(x, \pm, y)$. Such T.D.'s can produce geomagnetic impulses if n is
sufficiently large; those with signatures (x, o, y) cannot. (A small change in V_w may be
observed across a T.D. due to motions along the glide plane, but they do not produce
significant impulses.) Direct evidence for hydromagnetic tangential discontinuities
in the solar wind was presented by Burlaga (1968) and Burlaga and Ness (1969),

and current sheets characteristic of those at T.D.'s were identified by Siscoe *et al.* (1968a).

A *contact surface* is frequently confused with a T.D. since both are non-propagating and the pressure is continuous across both. But there is a fundamental difference: there is a component of **B** normal to a contact surface and $\mathbf{B_1} = \mathbf{B_2}$, whereas at a T.D.

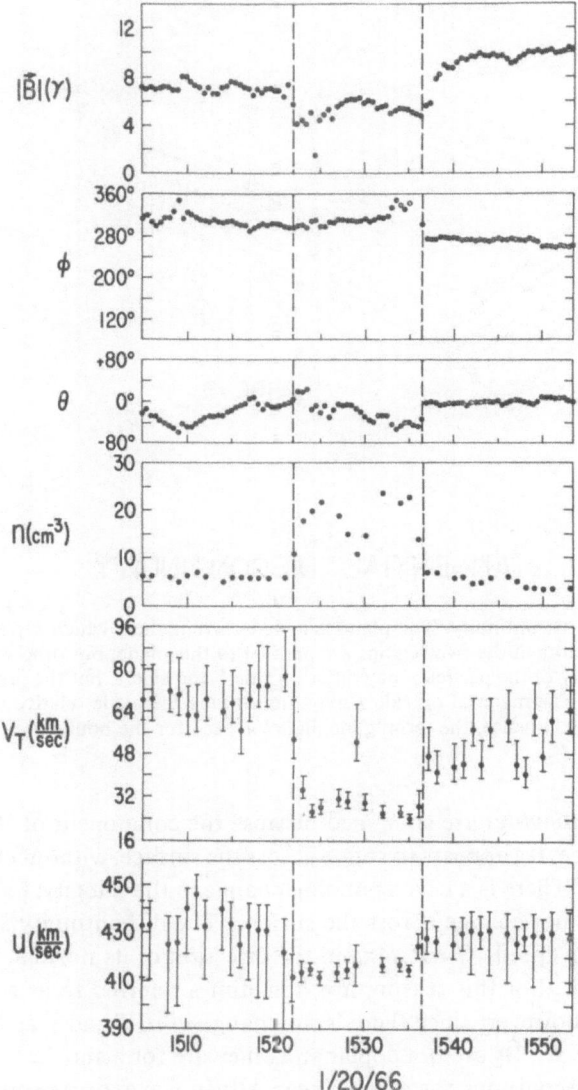

Fig. 1.   Two discontinuities in the interplanetary magnetic field. The abscissa is universal time. The ordinates give the magnetic field intensity |**B**| and direction ϕ, θ, in solar ecliptic coordinates, the plasma density n, the thermal speed V_T and the bulk speed u. In this case the two discontinuities appear to define a 'filament', but usually discontinuities are *not* paired.

B is parallel to the surface and in general $\mathbf{B}_1 \neq \mathbf{B}_2$. There can be no relative motions of the two regions separated by a contact surface, so $\mathbf{V}_{w1} = \mathbf{V}_{w2}$. Contact surfaces could give rise to geomagnetic impulses, since $n_2 \neq n_1$ (the pressure is balanced by a corresponding change in the temperature), but none has yet been identified in the solar wind. Two spacecraft are needed to distinguish a contact surface from a T.D. with signature $(0, \pm, \mp)$ and $\mathbf{B}_1 = \mathbf{B}_2$.

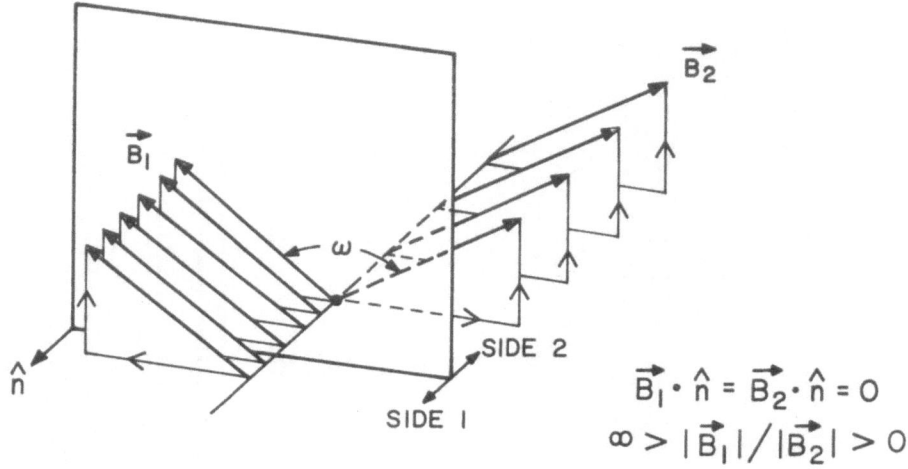

$$\vec{B}_1 \cdot \hat{n} = \vec{B}_2 \cdot \hat{n} = 0$$
$$\infty > |\vec{B}_1|/|\vec{B}_2| > 0$$

TANGENTIAL DISCONTINUITY

Fig. 2. A tangential discontinuity. The plane is a thin current sheet which separates two regions. The magnetic field vectors in the two regions are parallel to this plane, but otherwise arbitrary. The temperature and density of the particles may differ on side 1 and side 2, but the pressure must be the same in both regions. The material on side 1 may move along the plane relative to the material on side 1; hence, the term 'glide plane' is used for the boundary.

Rotational discontinuities are so named because the component of **B** tangent to the discontinuity surface, $\mathbf{B}_t$, appears to rotate across the surface, without changing magnitude (see Figure 3). There is a corresponding change in the velocity, $[\mathbf{V}_t] = [\mathbf{B}_t]/(4\pi\varrho)$, but B, n, and T do not change across the surface. The discontinuity surface actually moves at the Alfvén speed $V_\perp = B_\perp/4\pi\varrho$ in the direction of its normal. Thus there is a mass flux through it. For this reason, it is sometimes referred to as a kind of shock; this is misleading, however, since there is no change in n, T and v as in a shock, and the (unit) vectors $\hat{n}$, $\hat{\mathbf{B}}_1$, $\hat{\mathbf{B}}_2$ are not coplanar as they are for a shock. It is better to picture a rotational discontinuity as a non-linear Alfvén wave, or a propagating kink in the magnetic field. Since $V_t \ll V_w$ (the solar wind speed), and since $\Delta n = 0$, the change in momentum flux across an R.D. is small $\Delta(mnV) \approx \varrho\Delta V \approx \varrho V_w([V_t]/V_w)$ and is not likely to produce an observable geomagnetic impulse (see Section 4). Belcher *et al.* (1969)

presented solar wind observations which are consistent with R.D.'s, but they could also be interpreted as T.D.'s. It is very difficult to distinguish an R.D. from a T.D. of the type (0,0,0), even if the solar wind direction is accurately known. The relative number of R.D.'s and T.D.'s in the solar wind is thus not known, but it is likely that most discontinuities are tangential.

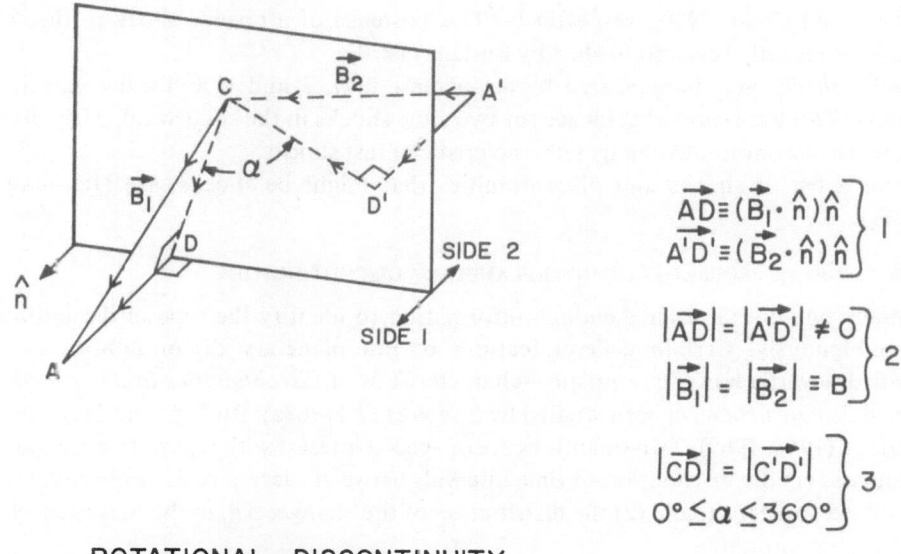

$$\vec{AD} \equiv (\vec{B_1} \cdot \hat{n})\hat{n} \atop \vec{A'D'} \equiv (\vec{B_2} \cdot \hat{n})\hat{n} \Bigg\} \ 1$$

$$|\vec{AD}| = |\vec{A'D'}| \neq 0 \atop |\vec{B_1}| = |\vec{B_2}| \equiv B \Bigg\} \ 2$$

$$|\vec{CD}| = |\vec{C'D'}| \atop 0° \leq a \leq 360° \Bigg\} \ 3$$

ROTATIONAL DISCONTINUITY

Fig. 3. A rotational discontinuity. The plane is an element of a real surface which can be measured in space. There is a component of **B** normal to the surface. The field intensity does not change across the surface; thus, the tangential component of **B** appears to rotate in the plane through the angle α. The density and temperature do not change across the plane. The plane of the discontinuity propagates relative to the plasma with the Alfvén speed.

Fast shocks are analogous to ordinary gas-dynamic shocks, the difference being that across a fast shock the magnetic field intensity increases as well as the density and temperature. Relative to the shock speed, the flow speed *decreases* across the shock, i.e., $v_b < v_f$ (b=behind the shock, f=ahead of it). When a fast shock propagates *away* from the Sun, as is usually the case in the solar wind, $v_f = U - V_f$ and $v_b = U - V_b$, where U and V are the shock speed and the solar wind speed, respectively, relative to the Sun; thus $v_b < v_f$, $V_b > V_f$. In other words the solar wind speed measured relative to a fixed frame appears to increase across a fast shock moving away from the Sun. Now, it is also possible for a fast shock to propagate *toward* the Sun, yet move *away* from it, if it propagates slower than the solar wind speed. (It is like a man trying to walk slowly up a downward moving escalator). In this case one first sees the flow behind the shock, so n, B and T appear to decrease with time. The speed always decreases behind the shock, so $v_f = U + V_f > v_b = U + V_b$ implies $V_f > V_b$. Since one first sees the flow behind this shock (V_b), the solar wind speed appears to *increase* with time. A fast shock propagating toward the Sun is called a 'reverse shock'. Summarizing, a fast shock moving outward has the signature $(+, +, +)$ and V increases, while a 'reverse fast shock' has the signature $(-, -, -)$ and V increases. The momentum flux relative to the Earth increases across a fast shock moving away from the Sun, and decreases across a 'reverse fast shock'. The existence of fast shocks in the solar wind has been established with increasing certainty by Sonett *et al.* (1964), Ogilvie and Burlaga (1969), and Chao (1970), respectively. The existence of a reverse shock in the solar wind has recently been established by Burlaga (1970).

Slow shocks are characterized by an increase in n, T and V and a decrease in B. Chao (1970) has reported evidence for two slow shocks in the solar wind. They do not show the discontinuous changes characteristic of fast shocks.

For a list of shocks and discontinuities that might be shocks, see Hundhausen (1969).

2.2. GENERAL PROPERTIES OF INTERPLANETARY DISCONTINUITIES

Usually one does not have enough information to identify the type of discontinuity unambiguously. Certain general features of interplanetary discontinuities can be studied nevertheless. The statistical characteristics of *discontinuities in the interplanetary magnetic field* have been studied by Siscoe *et al.* (1968a), Burlaga and Ness (1968), Burlaga (1968, 1969). Two quantities are of special interest with regard to geomagnetic impulses: (1) the distribution of time intervals between successive discontinuities seen at a fixed spacecraft, and (2) the distribution of the change, ΔB, in the magnitude of **B** across discontinuities.

Let ω be the change in the direction of **B** across discontinuities. Siscoe *et al.* (1968a) defined a discontinuity by the condition $|\mathbf{B}(t_2) - \mathbf{B}(t_1)| \geqslant 4\gamma$, which implies $\omega \gtrsim 20°$ for $|\mathbf{B}_1| = |\mathbf{B}_2| \approx 6\gamma$. Burlaga studied 'directional discontinuities' defined by $\omega > 30°$. Both assume that the change occurs in $\leqslant 1$ min. The two definitions are equivalent if $B_1 \approx B_2 \gtrsim 4\gamma$, as is usually the case.

Figure 4a shows the distribution of B_1/B_2 for 114 'planar' discontinuities (i.e. discontinuities which appear to be tangential) with thickness 10 sec $< t < 100$ sec from the Mariner 4 data for the period November 30, 1964 to January 3, 1965. Figure 4b shows the corresponding distribution of $(B_1 - B_2)/\text{Max}(B_1, B_2)$ for all the directional discontinuities from Pioneer 6 data for the period December 15, 1965 – January 1966. Although they are not strictly comparable, since Max $(B_1, B_2) = B_2$ only half of the time, the two distributions do show the same results; viz., (1) the magnitude, B, usually does not change significantly across a discontinuity, and (2) increases and decreases of B across the discontinuity are equally probable. The similarity between the Mariner and Pioneer results suggests that they refer to basically the same type of structures and indicates that the characteristics did not change appreciably during the year between the two measurements. Siscoe *et al.* (1968a) found that 'discontinuities' (actually current sheets), defined by $|B_2 - B_1| \geqslant 4\gamma$, occur roughly at the rate of one per hour at quiet times. This does not completely describe the time distribution, since there are many discontinuities with $|B_2 - B_1| < 4\gamma$ which occur even more frequently than 1/hr. Burlaga divided the discontinuities into four classes and obtained the distributions shown in Figure 5 for the time intervals between successive discontinuities in each class. The discontinuities with $30° < \omega < 60°$ occur with a mean separation $\lesssim 2$ hr. Smaller discontinuities ($\omega < 30°$) occur even more often, but are more difficult to identify and measure.

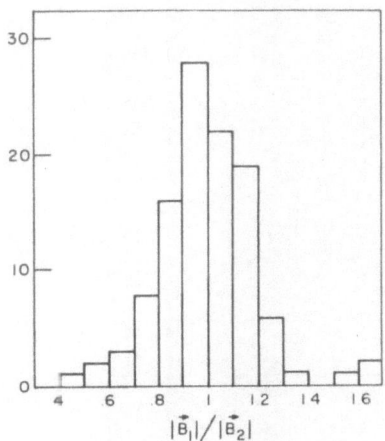
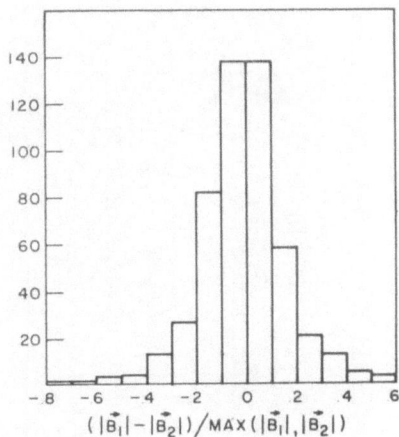

Fig. 4. The distribution of the change in magnitude of **B** across discontinuities in the solar wind. The distribution on the left (a) is based on Mariner 4 data and that on the right (b) is based on Pioneer 6 data obtained a year later. The two distributions are essentially the same. The field intensity usually does not change across a discontinuity, and the change is seldom $\geqslant 20\%$.

Unfortunately, there are as yet no distributions for the *discontinuities in plasma parameters*, n, V and T, corresponding to the magnetic field distributions described above. There are at least two reasons for this: (1) the plasma parameters are not measured as accurately as the magnetic field, and (2) the time between successive measurements is relatively long, usually 1–5 min, so that it is difficult to distinguish small discontinuities from continuous changes. It may also be found that the changes in plasma parameters are not as abrupt as changes in the magnetic field direction. Since the desired distributions are not likely to be forthcoming for some time, yet are of basic importance for studying small impulses in the Earth's magnetic field, we shall venture to make some order-of-magnitude estimates. Suppose that most discontinuities are tangential discontinuities, so that $B^2/8\pi + nk(T_p + T_e) = $ constant, and suppose that $(T_p + T_e) \approx$ constant for most discontinuities. Then, assuming $\beta = nk(T_p + T_e)/(B^2/8\pi) \approx 1$, $2(\Delta B/B) \approx \Delta n/n$, and a 40% change in n will be caused by a 20% change in B. Figure 4 shows that B changes by $\gtrsim 20\%$ for $\approx 25\%$ of the directional discontinuities. Since directional discontinuities occur at the rate of $\approx 1/(2$ hr), we expect density changes to occur at the rate of $\sim 1/(8$ hr), or 3/day, with increases and decreases being equally probable. There are times when density discontinuities may occur more frequently. For example, Siscoe *et al.* (1968a) and Burlaga (1969) showed a

series of density discontinuities following a shock which was apparently driven by a
high-speed stream (see Figure 6). Changes in the bulk speed may occur, but will probably be small ($\lesssim 5\%$) for most discontinuities. A study of large changes in the bulk
speed by Burlaga (1969) showed only six cases with $\Delta V > 60$ km/sec in ≈ 2500 hr of
data; thus, discontinuities with $\Delta V/V \gtrsim 15\%$ occur at roughly the rate $\approx 1/15$ days.
Such discontinuities are not important as regards geomagnetic impulses.

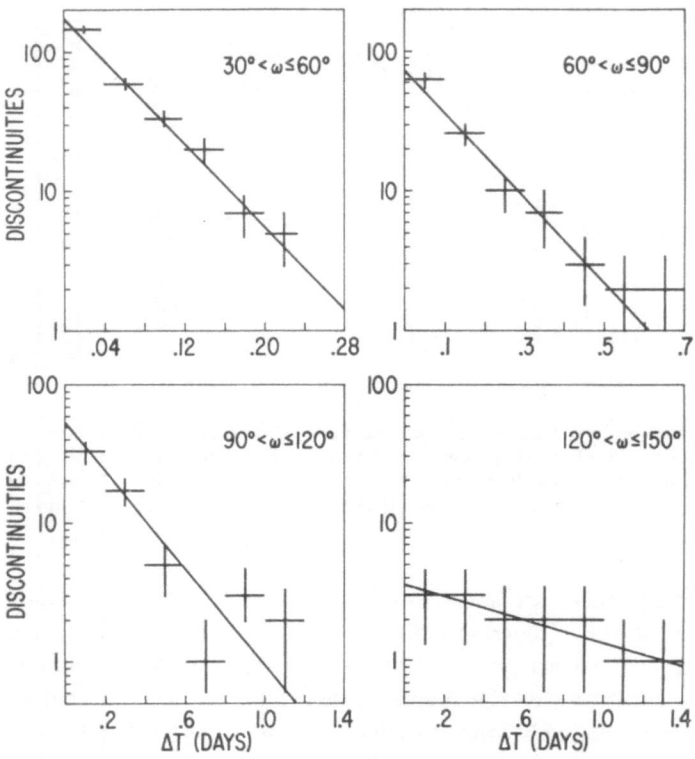

Fig. 5.   Distribution of time intervals between successive discontinuities, sorted in groups according
the ω, the change in the magnetic field direction. Discontinuities with small changes
occur most frequently.

2.3. FILAMENTS AND SECTOR BOUNDARIES

The discovery of numerous discontinuities in the magnetic field direction and magnitude (Ness et al., 1964, 1966) and in the direction of anisotropic cosmic ray fluxes
(Bartley et al., 1966) led to the suggestion that the interplanetary magnetic field could
be pictured as a bundle of corotating, intertwined, spaghetti-like 'tubes' or filaments
with sharp boundaries which extended from the Sun to the Earth's orbit and beyond
(McCracken and Ness, 1966). The diameters of these tubes was put at $(0.5–4) \times 10^6$
km. Additional support for this attractive picture was given by Siscoe et al. (1968a)

who suggested that the tubes are actually elliptical as the result of latitudinal solar wind shear. Hundhausen *et al.* (1967 a, b) reported discontinuous changes in the plasma parameters and interpreted this as boundaries of filaments with a scale size near 0.01 AU.

This picture of filaments grew out of preliminary work based on small data samples. Burlaga (1968) examined 500 hr of magnetic field data from Pioneer 6 and pointed out

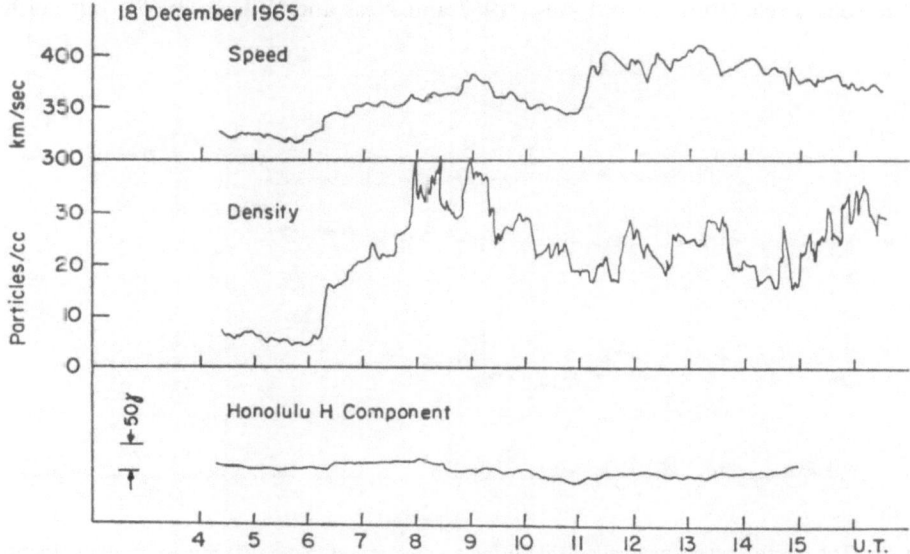

Fig. 6.   This shows a series of discontinuities in the density in the flow behind a shock at 0610. The magnetic field data shows a corresponding pattern. Filamentary forms can be seen, but they are not unambiguous and are not all bounded by directional discontinuities. This figure shows the material that is piled up by an advancing stream of fresh, hot plasma.

that discontinuities were always present and could be quantitatively defined and analyzed, but filaments could not always be recognized or defined. For example, Figure 7 shows a quiet day with 11 clearly defined directional discontinuities, but the identification of filaments would be very subjective. He also noted that in general there is no obvious pairing of discontinuities. Thus, Burlaga suggested that the solar wind should be regarded as discontinuous rather than filamentary and he pointed out that one should not discard the possiblity that discontinuities are created and destroyed in the interplanetary medium.

There are occasional times, however, when filamentary forms can be seen, particularly behind shocks (see Figures 1 and 6), but these forms are not always bounded by sharp discontinuities. The class of 'box-like' events discussed by Siscoe *et al.* (1968a) and those of Ness *et al.* (1966, Figure 8) might also be properly termed filaments. The behavior of isolated filaments has been investigated analytically by Siscoe (1970).

There is as yet no general, quantitative definition of a filament. Until one is given, it might be better not to speak of the radius of filaments or the topology of filamentary magnetic tubes.

The boundary between sectors (Ness *et al.* 1964 and Wilcox and Ness, 1965) is sometimes discontinuous. Several authors have given special geophysical significance to these boundaries but their importance is probably overestimated. Sector boundaries are not always discontinuous or well defined, particularly during the more active parts of the solar cycle (Burlaga and Ness, 1967 and Ness and Wilcox, 1967); but neither

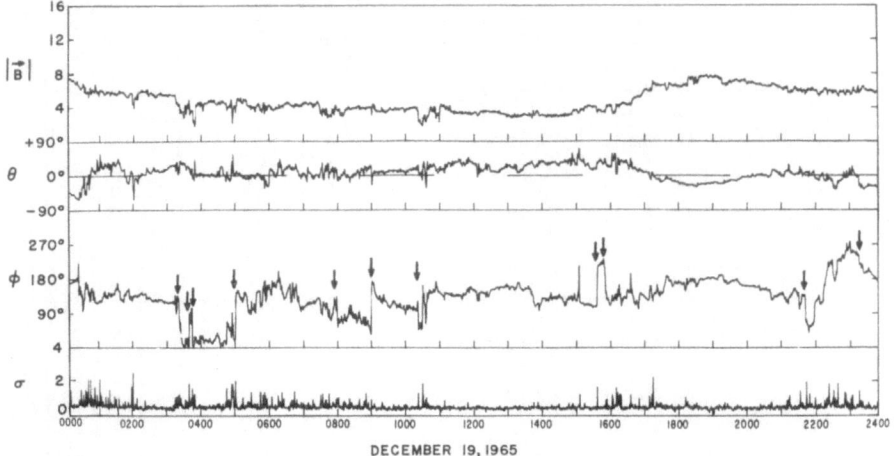

Fig. 7.   The interplanetary magnetic field during a quiet period. Note that numerous discontinuities can be seen (marked by arrows), but it is not possible to describe the field by a unique series of step functions. Thus, the term 'discontinuities' is appropriate, but it is an oversimplification to think of the field as a superposition of distinct filaments. However, some filamentary forms can be seen. The plot is based on 30-sec averages of the magnetic field; σ is the standard deviation for each of the averages.

are they 'turbulent'. When they are discontinuous, it is usually a directional discontinuity with no change in the magnitude of **B**, so there is generally no corresponding geomagnetic impulse. Nishida (1966a) discusses a positive 'sudden impulse' (not reported by geomagnetic observatories) associated with a sector boundary. This directional discontinuity was associated with a dip in the magnetic field intensity, and thus has the character of a '*D*-sheet'. Such structures were studied by Burlaga and Ness (1968) and Burlaga (1968) who find that they do not always occur at sector boundaries, and that they are accompanied by an increase in density; thus, they could produce a geomagnetic impulse as suggested by Nishida (1966a) but not all sector boundaries would give such an impulse, and such impulses may occur in the absence of sector boundaries.

3. Types of Geomagnetic Impulses

Several types of impulses are seen in ordinary magnetograms, usually most clearly in the H (horizontal) component of the Earth's magnetic field. The Provisional Atlas of Rapid Variations (1957) classifies the impulses as $si^{\pm}$, $ssc^{\pm}$, and ssc^*. These are illustrated in Figure 8 and defined as follows:

(1) $si^{\pm}$: (a) An abrupt increase $(+)$ or decrease $(-)$ in the magnetic field which is *not* followed by an appreciable increase in activity; (b) a small reversed impulse, not followed by an increase in activity; and (c) a large, distinctive impulse similar to (a) above except that it occurs during a storm.

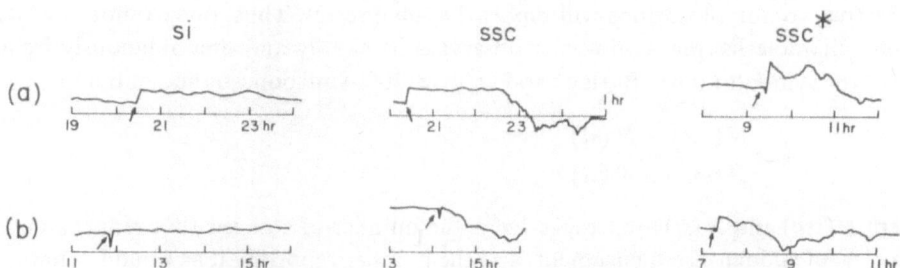

Fig. 8. Definition of geomagnetic impulses in the H component of the Earth's field. Events of type (a) are positive impulses. Corresponding events with a decrease in H, rather than the increase shown in (a) are negative impulses. Other types of negative impulses, shown by (b), are less frequently seen.

(2) $ssc^{\pm}$: (a) a sudden impulse (positive $+$, or negative $-$), followed by an increase in activity lasting at least one hour (the intense activity of the storm may appear immediately or it may be delayed a few hours); (b) a reverse impulse followed by an increase in activity.

(3) ssc^*: (a) an ssc which is preceded by one small reversed impulse or (b) preceded by many small oscillations.

These are the principal types of impulses reported by observatories following the IAGA Symposium on Rapid Magnetic Variations in April 1957. See the Provisional Atlas for further details and additional examples. Observatories were asked to evaluate their identification of an impulse by the letters A (very distinct), B (fair, ordinary, but unmistakable) and C (doubtful). This is useful when deciding how to classify an event using all of the world-wide data.

The distinction between ssc and si was first suggested by Chapman (see Ferraro *et al.*, 1951).

Other classifications of geomagnetic impulses have been proposed (see Matsushita 1960, p. 1425 for references). Of special interest is that of Matsushita (1962) who distinguishes three types of sudden commencements (^-SC, SC, and SC^-) and three completely analogous types of sudden impulses (^-SI, SI, SI^-). The superscripts refer to small impulses preceding or following the main impulse which are found to be dependent on latitude and time, the dependence being the same for sudden commencements and sudden impulses. In the literature concerning spacecraft data these secondary pulses are often ignored since they are due to ionospheric currents, and one frequently finds the symbols *SSC* or *SC* and *SI* denoting the two general classes of impulses distinguished by Matsushita. Matsushita distinguishes a fourth type of sudden impulse SI^θ which differs from SI^- in that it occurs simultaneously in the same form at all points on the Earth. He points out that there is no analogous SC^θ, an important point which is discussed in Section 4. The currents which give rise to the various types of sudden commencements are discussed by Sato (1961) and by Sastri and Jagakar (1967).

Clearly, the Atlas classifications depend on local time and latitude, and are also subjective, so not all stations will report the same result. Thus, one cannot in general simply characterize the world-wide observations simply and unambiguously by any one of the symbols above. Burlaga and Ogilvie (1969) introduced the symbol

$$A \equiv \frac{N(ssc) - N(si)}{N(ssc) + N(si)},$$

where $N(ssc)$ and $N(si)$ are, respectively, the number of stations that report an event as a type of sudden commencement, and the number reporting it as a sudden impulse. For events which according to *Solar Geophysical Data* were classified as a sudden commencement or sudden impulse by 10 or more observatories and occurred in the interval June-December 1967, Burlaga and Ogilvie (1969) found the distribution shown in Figure 9. Clearly, there are two classes, corresponding to sudden com-

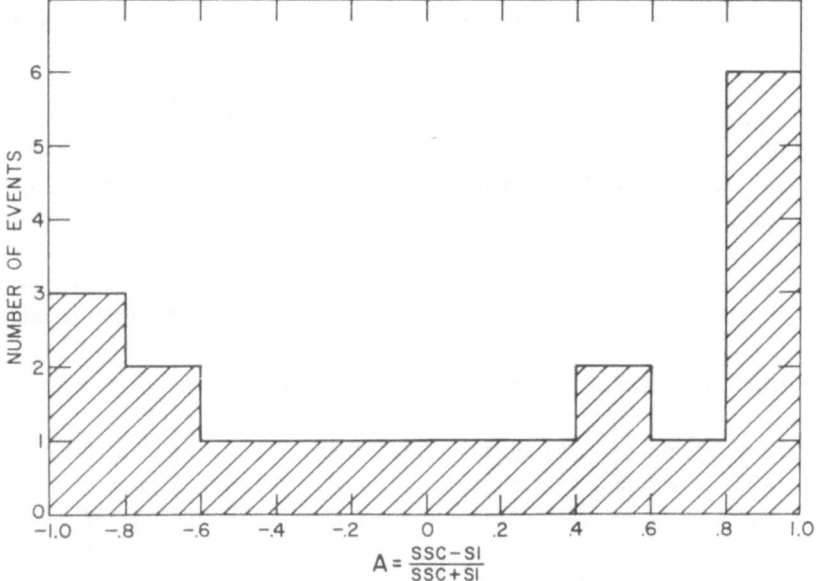

Fig. 9.   Not all observatories classify an event in the same way. The number A describes the relative number of stations that identify an event as a sudden commencement or a sudden impulse. There are two classes of events, corresponding to *si* ($A = -1$) and *ssc* ($A = +1$), but many events cannot be unambiguously classified as *ssc* or *si*. Apparently *ssc*'s are more easily recognized than *si*'s.

mencements ($A \geqslant 0.8$) and sudden impulses ($A \leqslant -0.8$), but *there are also many events that do not fall into these classes*. A further complication has been pointed out by Oguti (1968) who notes that several discontinuities may occur between the initial impulse of a storm and the main phase and the largest of these will be selected as the *ssc*, thus possibly giving $A > 0.8$ when the event might more appropriately be denoted as *si*. Thus, when space observations are related to ground-observed impulses, care should be used in characterizing the impulse.

Bowling and Wilson (1965) presented a collection of observational results, showing that *ssc*'s and *si*'s have 10 characteristics in common. They infer, as did others previously, that *ssc* and *si* are essentially the same phenomenon, both being caused by a sudden compression of the magnetosphere as the result of a discontinuous change of the energy density in the solar wind.

Nishida and Jacobs (1962) showed that there are other rapid 'world-wide changes' in the geomagnetic field which are not reported as sudden commencements or sudden impulses, yet have the same form, manner of spreading over the earth, and distribution of magnitude as *ssc*'s and *si*'s. They are more similar to *si*'s than *ssc*'s in that they are usually not followed by increased magnetic activity. Nishida and Jacobs suggested that *si*'s are nothing more than world-wide changes that are widely recognized because of their large size. Both positive and negative world-wide changes are observed with essentially the same probability. At least 90% of the days and at least 20% of all 1-hr periods that they examined contained one or more world-wide changes.

4. Relations Between Types of Geomagnetic Impulses and Interplanetary Discontinuities

4.1. RELATIONS BETWEEN TYPES

Let us define *ssc* by $A > 0.8$ and *si* by $A < -0.8$, and ask whether there is a relation between the type of impulse ($ssc^{\pm}$, $si^{\pm}$) and a particular type of discontinuity.

ssc$^+$ and shocks. Evidence for hydromagnetic shocks that caused an ssc^+ was presented by Sonett *et al.* (1964), Dryer and Jones (1968), Burlaga and Ogilvie (1969) and Chao (1970). Conversely, Burlaga and Ogilvie (1968) showed that ssc^+ is a fairly reliable indication of a hydromagnetic shock. Taylor (1968), using only interplanetary magnetic field data, examined the causes of 36 events reported as *ssc* by most stations ($A > 0.5$) during 1965, 1966, and 1967. He found that (1) 26 of these were likely to be caused by shocks, and (2) 10 were *not* caused by shocks, 5 of which had $A > 0.8$. Thus a sudden commencement is a very good indication of a shock, but there are exceptions.

ssc$^+$ and T.D.'s. An example of a sudden commencement that was *not* caused by a shock is given by Taylor (1968) – an event classified as *ssc* and *si* by 42 and 3 stations, respectively. It was associated with a large decrease in B and a $>90°$ change in the direction of B, and the positive geomagnetic impulse implies an increase in density; thus, the discontinuity must be tangential, with signature $(-, +, ?)$. Such events are relatively rare, however. It should be noted that in this case the main phase immediately followed the impulse; thus it represents the type of storm which Oguti (1968) attributes to a 'bubble' (driver gas) that is not preceded by a shock.

Gosling *et al.* (1967a) reported that a world-wide 'sudden commencement' at 1223 UT on April 6, 1965 was associated with a discontinuity whose signature was $(?, ?, -)$; they note that this could be a tangential discontinuity, and is not a fast shock. Lincoln (1966) shows that 26 stations identified the event as *ssc* with 'distinctness indices' $(A:4, B:14, C:8)$, and 14 identified it as *si* $(A:4, B:5, C:5)$, giving $A = 0.3$; the event is not clearly a sudden commencement.

ssc⁻. As discussed above, it is not clear that there is such a thing as a worked-wide *ssc*⁻. In any case, as Oguti (1968) pointed out, it would be very difficult to distinguish it from an *si*⁻ that just happens to occur during a storm, particularly if the negative impulse happens to be larger than any associated positive impulse. Gosling *et al.* (1968) showed that an interplanetary discontinuity with signature $(?, -, +)$ caused a negative impulse which they identified as a 'relatively rare negative *SC*'. However, this event was classified as *ssc*⁻ by 16 stations and as *si*⁻ by 15 stations (Lincoln, 1965) which gives $A = 0.03$, so that we would not call it a sudden commencement. Akasofu (1964) discusses an event at 0718 UT on 10 January, 1960 that was classified *ssc*⁻ by 44 stations and *si*⁻ by 12 stations, which gives $A = 0.39$. This is a case in which two world-wide impulses occurred. The positive impulse at 0610 UT was not reported, but the larger negative impulse was identified and associated with the geomagnetic activity that followed. Most reported instances of *ssc*⁻ are probably in this class, and could equally well be described as *si*⁻ which occurred after the positive impulse of an *ssc*⁺. There appear to be no *ssc*⁻ analogous to *ssc*⁺, with $A > 0.8$. In any case, there is no evidence for an *ssc*⁻, even with $A > 0.5$, that is caused by a 'reverse fast shock'. The reverse shock identified by Burlaga (1970) was not associated with a geomagnetic impulse.

si⁺. There are few reported observations of interplanetary observations associated with sudden impulses. Burlaga and Ogilvie (1969) show cases in which *si*⁻ was caused by a tangential discontinuity across which the density decreased. There is no evidence supporting the suggestion of Sonett and Colburn (1965) that *si*⁻ is caused by a reverse shock. Gosling *et al.* (1967b) show a $(?, +, +)$ discontinuity at an *si*⁺.

'*World-wide impulses*'. The causes of these have not been extensively studied, but they are probably due to tangential discontinuities. Gosling *et al.* (1967a) showed plasma data for two world-wide impulses, (one of them negative) which were not reported as *si* (see Figure 10). The signatures, $(?, +, -)$ and $(?, -, 0)$, for the positive and negative impulse, respectively, clearly exclude shocks and suggest T.D.'s as the causes. Since a density change is usually opposite to the magnetic field change across a T.D. and since positive and negative changes in B are equally probable (see Section 2), one should observe equal numbers of positive and negative world-wide impulses, in agreement with the observations (see Section 3). From (1) below and Figure 11, we see that a change $\Delta\sqrt{P} \gtrsim 10^{-4}$ (dynes/km^2)$^{1/2}$ which corresponds to $\Delta n/n \approx 2\sqrt{n} \cdot \Delta\sqrt{n}/n$ $\gtrsim 0.4$, will produce an observable impulse $(\Delta H \gtrsim 10\gamma)$. This in turn implies $\Delta B/B \approx$ $\approx \Delta n/2n \gtrsim 0.2$, and it was shown that such discontinuities occur at the rate of $\approx 3/$day,

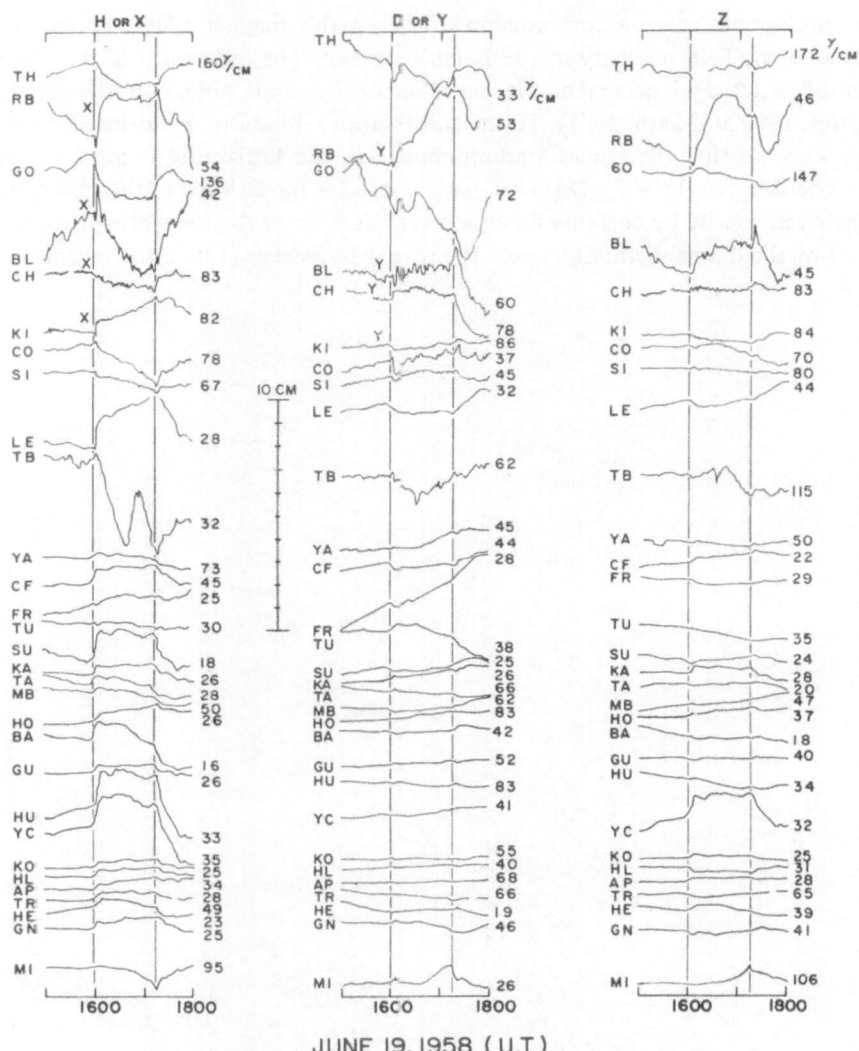

JUNE 19, 1958 (U.T.)

Fig. 10.   The magnetogram traces show impulses that occurred world-wide, but were not identified as *si* or *ssc* by geomagnetic observatories. The corresponding plasma data suggests that these 'world-wide impulses' were caused by the density changes at tangential discontinuities.

which is on the order of the rate of occurrence of world-wide impulses (Section 3). Thus, it seems likely that world-wide impulses are due to T.D.'s. This supports the suggestion of Nishida and Jacobs (1962) that world-wide impulses are the same as *si*'s.

4.2. RELATION BETWEEN THE SIZE OF THE IMPULSES AND THE INTERPLANETARY DISCONTINUITY

Parker (1958) pointed out that a discontinuous increase in the momentum flux of the solar wind would cause a compression of the Earth's magnetic field which gives an increase in the field intensity at the Earth's surface. The induced field has the same magnitude, ΔB, and occurs nearly simultaneously at all points of the Earth (see Williams, 1960 and Sato, 1961). To the zeroth approximation, it is oriented along the dipole axis, so that the corresponding change in the horizontal component of the Earth's field is given by $\Delta H_{obs}(\lambda) = \Delta B \cos \lambda$ where λ is the latitude of the observer. It is generated primarily by currents flowing on the surface of the magnetosphere and enhanced by the diamagnetic Earth (see the review by Parker (1962), the recent analysis

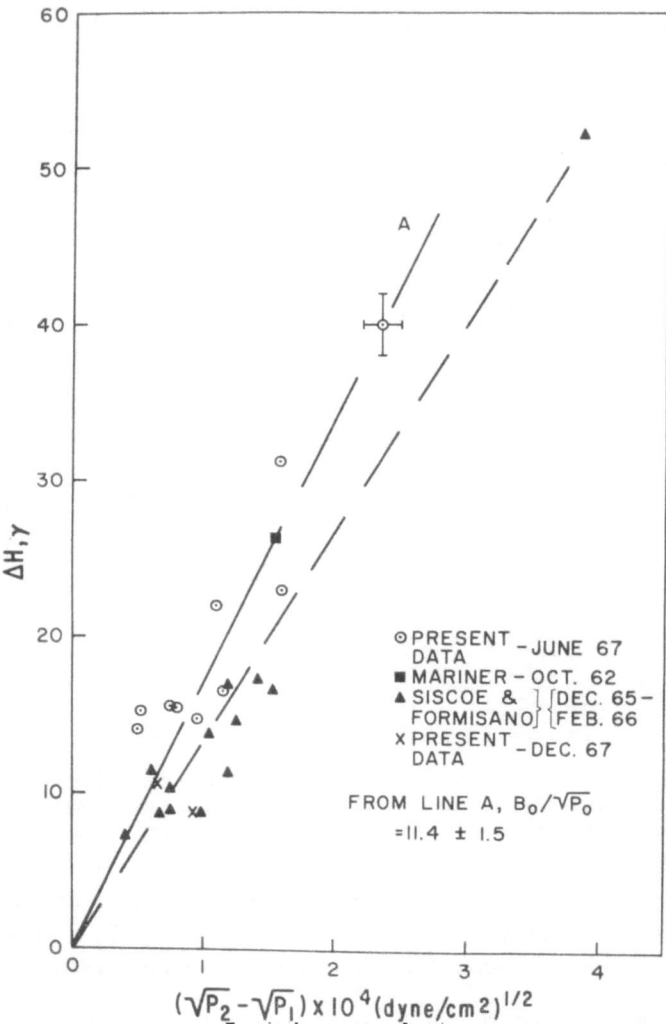

Fig. 11.   This shows that the change in the horizontal component of the Earth's field (divided by $\cos \lambda$) is proportional to the change in the momentum flux across a discontinuity, as predicted. The difference between the two sets of observations might be a seasonal effect; this requires further study.

by Siscoe (1966), and the pioneering paper by Chapman and Ferraro (1931)). Siscoe *et al.* (1968b) give the following expression for ΔB, from the magnetosphere model of Mead (1969):

$$\Delta B = A(\sqrt{P_2} - \sqrt{P_1}), \tag{1}$$

where $A = 26.1 \times 10^4 \gamma/(\text{dyne/cm}^2)^{1/2}$ and, $P = b \times 1.16 \, m_p n_p V_w^2$, where b ranges from 0.88 for $\gamma = \frac{5}{3}$ to 0.955 for $\gamma = 1.2$ ($\gamma \equiv$ adiabatic exponent), m_p and $n_p \equiv$ proton mass and density, $V_w \equiv$ bulk speed, and the helium density is taken to be 4% of the proton density to give $1.5 \, m_p n_p$ for the total density. Using Mariner plasma data for 13 discontinuities associated with *si*'s and world-wide impulses in the period December 1965-February 1966, Siscoe *et al.* (1968b) found that $\Delta B \approx A \Delta \sqrt{P}$ where $A \approx (9.0 \pm 2.) \times 10^4 \gamma/(\text{dynes/cm}^2)^{1/2}$. Similarly, Ogilvie *et al.* (1968) used plasma data from Explorer 34 for discontinuities associated with *si*'s and *ssc*'s during June 1967 to show that $\Delta B \approx A \Delta \sqrt{P}$ where $A = (11.4 \pm 1.5) \times 10^4$. Thus the linear relation given by (1) is confirmed. But the experimental value of A is less than $\frac{1}{2}$ the theoretical value; Siscoe (1970) suggests that this may be due in part to the presence of magnetospheric particles. The difference between the two experimental values of A is small ($\lesssim 20\%$) but might be real.

Ogilvie *et al.* did not consider that the size of the *ssc* impulse is enhanced on the day side of the Earth at geomagnetic latitudes $\lesssim 20°$ by ionospheric currents (see Sugiura, 1953; Jacobs and Watanabe, 1963; Rastogi *et al.*, 1966; Srinivasmorthy, 1960; and Maeda and Yamamoto, 1960). Correcting their work for this effect gives no significant change in the results in Figure 11.

Typically, ΔH is on the order of 30γ or 40γ for *ssc*, but Bhargava and Natarajan (1967) describe an event on November 13, 1960 with $\Delta H = 368\gamma$ at Trivandrum, 220γ at Kakioka and 211γ at Alibag. Using the lower values and Equation (1), we find $\Delta(nV_w) \sim 1.5 \times 10^9$. For a strong shock with $n_2 = 4n_1$, for the extreme case $V_2 \approx 3V_1 \approx \approx 900$ km/sec, this implies $n_1 \approx 10^2 \text{cm}^{-2}$, which is a very high density. They also note the events of July 17, 1959 and July 11, 1959 for which $\Delta H \sim 127\gamma$ and 102γ, respectively, seen at night at Trivandrum.

Further studies of the relation (1) should be undertaken to better understand the currents which relate the surface effect to the interplanetary cause. Once the Earth's field is thus calibrated, one can analyze the interplanetary discontinuities which left their imprint on magnetograms in the pre-satellite era.

Rise Time. The change, ΔH, in a geomagnetic impulse occurs over a relatively long time interval, $\approx 1-5$ min, which is called the rise time. There are at least three explanations for this: (1) Nishida (1964, 1965b) suggested that the rise time for *ssc* is determined by the nature of the interplanetary discontinuity; (2) Dessler *et al.*, (1960) and Francis *et al.* (1959) suggested that it is determined by the time it takes hydromagnetic waves to propagate through the magnetosphere from the various parts of the surface of the magnetosphere; (3) Sugiura suggested that it is determined by the transition from the initial state to the final state in the outermost region of the magnetosphere.

Nishida (1964) related rise time of *ssc* to an indirect determination of $V_r = V - V_w$, the mean speed of the discontinuity between the Sun and the Earth relative to the solar wind speed. He distinguished between two kinds of *ssc*: those with $V_r \gtrsim 600$ km/sec, which were associated with short rise times ($\lesssim 2$ min), and those with $V_r \lesssim 600$ km/sec, which were associated with longer rise times ($\gtrsim 2$ min). He attributed the former to shocks, the latter to discontinuities of some non-shock mode. If we assume that his *ssc*'s correspond to $A \gtrsim 0.8$, there is clearly a disagreement between his inference and the conclusion in Section 4 concerning the general cause of *ssc*. There is another problem: there is no hydromagnetic discontinuity other than a shock which propagates with speeds 100–500 km/sec relative to the solar wind, as does Nishida's non-shock mode. Finally, Nishida assumed that the rise time is determined primarily by the thickness of the interplanetary discontinuity, but some evidence by Burlaga and Ogilvie (1968) argues against this.

The quantitative theory of Dessler *et al.* (1960) is two-dimensional and based on the geometrical ray approximation. Stengelman and Kenschitzki (1964) extended the theory using a three-dimensional model and found that it could not explain the shape of the impulse, because of the inadequacy of the ray approximation.

Sugiura's explanation implies that the rise time should be essentially the same everywhere in the magnetosphere, in agreement with the Explorer 12 results of Nishida and Cahill (1964). The faster the shock, the shorter the transition time (see the illustrative calculation of Spreiter and Summers, 1965), and thus the shorter the rise time. A faster shock is also a stronger shock for a given solar wind speed, which implies a larger momentum change ΔP and thus a larger impulse ΔH. One concludes, then, that the theory implies an inverse relation between the rise time and ΔH for *ssc*. Pisharoty and Srivastava (1962) showed that such a relation does exist for the *ssc*'s at Alibag between 1949 and 1960 (Figure 12). Chapman and Bartels (1962, p. 297) suggest no such relation, however, on the basis of the points shown as ×'s in Figure 12.

To really understand the rise time a good model of the propagation of an impulsive disturbance through the magnetosphere is needed. There are many models (e.g. Hines and Storey, 1958; Hines, 1958; Ferraro *et al.*, 1966; and Willis, 1964) but we shall not pursue the subject.

4.3. 'SI⁺ – SI⁻ Pairs'

Geomagnetic Observations. Sugiura *et al.* (1965) pointed out that world-wide impulses often occur in pairs consisting of a small positive impulse with $\Delta H \sim 5\gamma$, followed approximately one hour later by a similar negative impulse. Figure 13 shows an example of such a pair.

Akasofu (1964) pointed out the existence of another type of pair of impulses characterized by a large ($\sim 40\gamma$) positive impulse (not necessarily *si*⁺), followed several hours later by a similar negative impulse. He showed four such pairs, each of which was followed by geomagnetic activity.

A typical giant pair is shown in Figure 14. Sonett and Colburn (1965) introduced the term '*SI*⁺ – *SI*⁻ Pair' to describe both kinds of impulse pairs, but they distinguished between 'giant pairs' and 'regular pairs'. The term $SI^+ - SI^-$ pair is quite misleading and is best not used at all. The distinction between giant and regular pairs is, however, sound and useful.

Causes of Giant Pairs. Sonett and Colburn (1965) suggested that giant pairs are caused by a pair of convected shocks. One of these is an ordinary fast shock moving away from the Sun and causes the positive impulse; the other, which causes the nega-

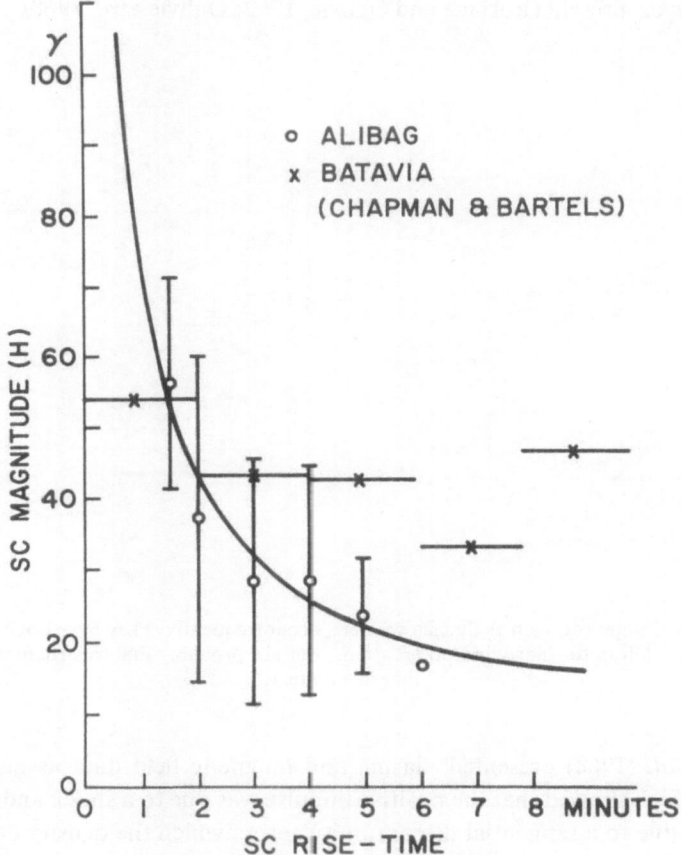

Fig. 12.   The change ΔH for *ssc* versus rise time, for data from Alibag (circles) and Batavia (crosses). Both an inverse relation (circles) and no relation (crosses) have been suggested.

tive impulse, is a reverse fast shock. The theory of such pairs was developed by several authors (Simon and Axford, 1966; Sturrock and Spreiter, 1965; and Schubert and Cummings, 1967, 1969). The most extensive model is that of Hundhausen and Gentry, 1969, who concluded that flare-associated forward-reverse shock pairs at 1 AU are not likely. Dessler and Fejer (1963) speculated that such shock pairs might appear at corotating streams in the solar wind. Razdan *et al.* (1965) noted the apparent

≈ 27-day recurrence of giant pairs in magnetograms and extended the Dessler-Fejer speculation to explain them.

Despite the extensive theoretical work, there is *no* direct evidence for a shock pair in the solar wind. Schubert and Cummings (1967, 1969) suggested that the shock seen on October 8, 1962 was one of a pair, but the contact discontinuity and trailing shock were *not* seen, and the features which they somewhat arbitrarily fit to the shock-pair model are very similar to those that are commonly seen in high-speed streams even when no shock is present (Burlaga and Ogilvie, 1969; Ogilvie *et al.* 1968).

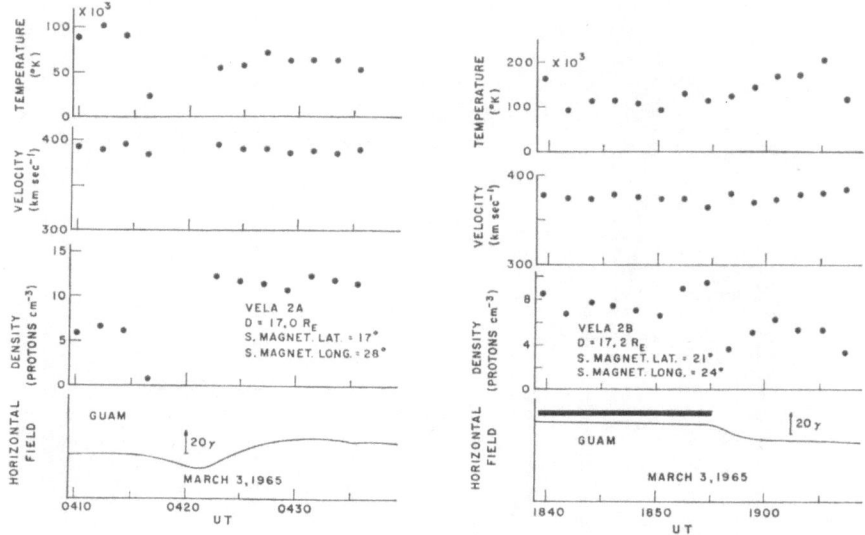

Fig. 13.   Pairs of impulses, such as that shown here, occur frequently. They have not been adequately studied in relation to solar wind observations, but are probably due to structures such as that in Figure 1.

Ogilvie *et al.* (1968) presented plasma and magnetic field data associated with a giant pair, which showed that the positive impulse was due to a shock and the negative impulse was due to a tangential discontinuity across which the density decreased (see Figure 14). Such a combination – a shock followed several hours later by an abrupt decrease in density – is frequently seen proceeding high-speed streams (for examples, see Burlaga and Ogilvie (1969); Ogilvie *et al.* (1968) and Lazarus *et al.* (1970). This suggests that giant pairs are generally caused by such driven shocks, the positive impulse being due to the shock as in an ordinary sudden commencement, and the negative impulse being due to the discontinuity which sometimes separates the driver gas from the high-density material that is piled up by the advancing stream. This discontinuity, then, would be analogous to that postulated by Parker (1963). The variety of possible giant pairs is reflected in the storm classification scheme of Oguti (1968).

The flow behind shock is usually not so simple, however, as evidenced by Figure 6, for example. Thus, the giant pair is just one of many types of geomagnetic signatures that can be produced by driven shocks, and the identification of the transition to driver gas is not always possible. The existence of several discontinuities behind certain shocks would be expected to be seen as *si* activity in the magnetograms. Yoshida and Akasofu (1966) have studied such events and related them to Forbush decreases.

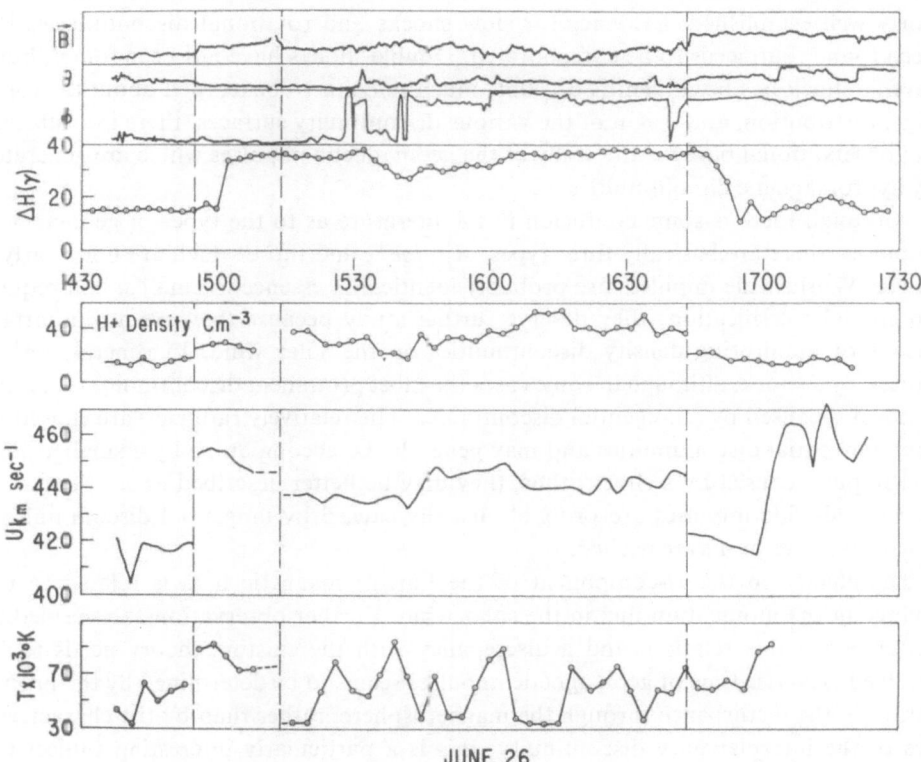

Fig. 14. The magnetogram trace shows a 'giant pair'. The corresponding plasma and magnetic field data show that the giant pair was caused by a shock followed by a tangential discontinuity. Such pairs may frequently appear with shocks driven by fresh, hot plasma from the Sun. The high-density material is probably compressed by the advancing stream; hence the second discontinuity of the pair may represent the transition between the driven and the driving gases.

There are available many unpublished observations of the flow behind shocks. Because of their complexity, an analysis of them should be based on a collection which is as complete as possible.

Causes of Regular Pairs. The relation between regular pairs and interplanetary observations has not been studied. Burlaga and Ogilvie (1969) showed two dense spots in the solar wind associated with geomagnetic pulses for which they suggested the symbol *pl*. Such pulses may simply be closely separated pairs. Figure 1 (from Burlaga,

1968) shows a complementary pair of tangential discontinuities between which the density is high; such a feature may be expected to produce a regular pair. This is clearly an area of special interest, which requires further study.

5. Summary

The existence of fast shocks and tangential discontinuities in the solar wind is now fairly well established. Evidence for slow shocks and rotational discontinuities has been found, but needs to be corroborated. Double shocks have not been found, but a reverse shock has been identified. While much remains to be learned about the topology, distribution, and origin of the various discontinuity surfaces, there is a substantial observational base for the study of the geomagnetic impulses which are generated by hydromagnetic discontinuities.

Although there is some confusion in the literature as to the types of geomagnetic impulses, there are basically three types: $si^{\pm}$, ssc^{+}, and those which are not clearly si or ssc. World-wide impulses are probably identical in essence to small si and require no special classification; they deserve further study because they are an important means of monitoring density discontinuities on the solar wind. In general, ssc^{+} is caused by a shock, although in some cases the most prominent discontinuity preceding a storm is caused by a tangential discontinuity. The relatively rare ssc^{-} are caused by such tangential discontinuities and may generally be accompanied by a smaller positive impulse caused by a shock; thus, they may be better described as si^{-}. The si^{-}'s and world-wide impulses are probably usually caused by tangential discontinuities. Further studies of si's are needed.

The change in the H-component of the Earth's magnetic field is related to the change in the momentum flux in the solar wind. Further observations are needed to better define this relation, and a discrepancy with the existing theory needs to be resolved. The rise time of geomagnetic impulses seems to be determined by the propagation of the disturbance through the magnetosphere, rather than by the characteristics of the interplanetary discontinuity; this is a particularly interesting subject for further study.

The subject of pairs of discontinuities ('$si^{+} - si^{-}$ pairs') is extensive but confused. It is suggested that giant pairs are usually caused by a shock followed by a tangential discontinuity, while regular pairs are usually due to complementary tangential discontinuities.

Acknowledgements

Drs. Ness, Ogilvie, Siscoe and Sugiura contributed several helpful comments on the manuscript.

References

(Includes several papers recommended by the author, but not cited.)

Akasofu, S. I.: 1964, *Planetary Space Sci.* **12**, 573.
Bartley, W. C., Bukata, R. P., McCracken, K. G., and Rao, U.R.: 1966, *J. Geophys. Res.* **71**, 3297.

Belcher, J. W., Davis, Jr., L., and Smith, E. J.: 1969, *J. Geophys. Res.* **74**, 2302.
Bhargava, B. N. and Natarajan, R.: 1967, *J. Atmospheric Terrest. Phys.* **29**, 957.
Bowling, Sue Ann and Wilson, C. R.: 1965, *J. Geophys. Res.* **70**, 191.
Burlaga, L. F.: 1968, *Solar Phys.* **4**, 67.
Burlaga, L. F.: 1969, *Solar Phys.* **7**, 72.
Burlaga, L. F.: 1970, NASA-GSFC preprint, X-692-70-95.
Burlaga, L. F. and Ness, N. F.: 1967, NASA-GSFC X-612-67-278.
Burlaga, L. F. and Ness, N. F.: 1968, *Can. J. Phys.* **46**, S962.
Burlaga, L. F. and Ness, N. F.: 1969, *Solar Phys.* **9**, 467.
Burlaga, L. F. and Ogilvie, K. W.: 1969, *J. Geophys. Res.* **74**, 2815.
Burlaga, L. F. and Ogilvie, K. W.: 1970, *Astrophys. J.* **159**, 659.
Chao, J. K.: 1970, Interplanetary Collisionless Shock Waves, CSR TR-70-3.
Chapman, S. and Bartels, J.: 1962, *Geomagnetism, Vol. II*, Clarendon Press, Oxford.
Chapman, S. and Ferraro, V. C. A.: 1931, *Terrest. Magn. Atmosph. Elect.* **36**, 171.
Colburn, D. S. and Sonett, C. P.: 1966, *Space Sci. Rev.* **5**, 439.
Dessler, A. J. and Fejer, J. A.: 1963, *Planetary Space Sci.* **11**, 505.
Dessler, A. J., Francis, W. E., and Parker, E. N.: 1960, *J. Geophys. Res.* **65**, 2715.
Dryer, M. and Jones, D. L.: 1968, *J. Geophys. Res.* **73**, 4875.
Francis, W. E., Green, M. I., and Dessler, A. J.: 1959, *J. Geophys. Res.* **64**, 1643.
Ferraro, V. C. A., Parkinson, W. C., and Unthank, H. W.: 1951, *J. Geophys. Res.* **56**, 177.
Ferraro, V. C. A. and Plumpton, C.: 1966, *An Introduction to Magneto-Fluid Dynamics*, Clarendon Press, Oxford.
Gosling, J. T., Asbridge, J. R., Bame, S. J., Hundhausen, A. J., and Strong, I. B.: 1967a, *J. Geophys. Res.* **72**, 3357.
Gosling, J. T., Asbridge, J. R., Bame, S. J., Hundhausen, A. J., and Strong, I. B.: 1967b, *J. Geophys. Res.* **72**, 1813.
Gosling J. T., Asbridge, J. R., Bame, S. J., Hundhausen, A. J., and Strong, I. B.: 1968, *J. Geophys. Res.* **73**, 43.
Hines, C. O.: 1958, *J. Geophys. Res.* **62**, 443.
Hines, C. O. and Storey, L. R. O.: 1958, *J. Geophys. Res.* **63**, 671.
Hundhausen, A. J.: 1969, in V. Manno and D. E. Page (eds.), *Intercorrelated Satellite Observations Related to Solar Events*, D. Reidel Publ. Co., Dordrecht, Holland, p. 111.
Hundhausen, A. J. and Gentry, R. A.: 1969, *J. Geophys. Res.* **74**, 2908.
Hundhausen, A. J., Asbridge, J. R., Bame, S. J., and Strong, I. B.: 1967a, *J. Geophys. Res.* **72**, 1979.
Hundhausen, A. J., Bame, S. J., and Ness, N. F.: 1967b, *J. Geophys. Res.* **72**, 5265.
Jacobs, S. A. and Watanabe, T. J.: 1963, *J. Atmospheric. Terrest. Phys.* **25**, 267.
Landau, L. D. and Lifshitz, E. M.: 1960, *Electrodynamics of Continuous Media*, Pergamon Press, London.
Lazarus, A. J., Ogilvie, K. W., and Burlaga, L. F.: 1970, *Solar Phys.* **13**, 232.
Lincoln, Virginia: 1965, *J. Geophys. Res.* **70**, 4963.
Lincoln, Virginia: 1966, *J. Geophys. Res.* **71**, 1477.
Maeda, H. and Yamamoto, M.: 1960, *J. Geophys. Res.* **65**, 2538.
Matsushita, S.: 1960, *J. Geophys. Res.* **65**, 1423.
Matsushita, S.: 1962, *J. Geophys. Res.* **67**, 3753.
McCracken, K. G. and Ness, N. F.: 1966, *J. Geophys. Res.* **71**, 3315.
Mead, G.: 1969, *J. Geophys. Res.* **69**, 1181.
Namikawa, T., Kitamuru, T., Okuzawa, T., and Araki, T.: 1964, *Rep. Ionosphere Space Res. Japan* **18**, 218.
Ness, N. F. and Wilcox, J. M.: 1967, *Solar Phys.* **2**, 351.
Ness, N. F., Scearce, C. S., and Seek, J. B.: 1964, *J. Geophys. Res.* **69**, 3531.
Ness, N. F., Scearce, C. S., and Cantarano, S.: 1966, *J. Geophys. Res.* **71**, 3305.
Nishida, A.: 1964, *Rep. Ionosphere Space Res. Japan* **18**, 295.
Nishida, A.: 1966a, *Rep. Ionosphere Res. Japan* **20**, 36.
Nishida, A.: 1966b, *Rep. Ionosphere Res. Japan* **20**, 42.
Nishida, A. and Cahill, L. J.: 1964, *J. Geophys. Res.* **69**, 2243.
Nishida, A. and Jacobs, J. A.: 1962, *J. Geophys. Res.* **67**, 525.
Obayashi, T. and Jacobs, J. A.: 1967, *J. Geophys. Res.* **62**, 589.

Ogilvie, K. W. and Burlaga, L. F.: 1969, *Solar Phys.* **8**, 422.
Ogilvie, K. W., Burlaga, L. F., and Wilkerson, T. D.: 1968, *J. Geophys. Res.* **73**, 6809.
Oguti, T.: 1968, *Rep. Ionosphere Space Res. Japan* **22**, 37.
Parker, E. N.: 1958, *Phys. Fluids* **1**, 171.
Parker, E. N.: 1962, *Space Sci. Rev.* **1**, 62.
Parker, E. N.: 1963, *Interplanetary Dynamical Processes*, Interscience, New York.
Pisharoty, P. R. and Srivastava, B. J.: 1962, *J. Geophys. Res.* **67**, 2189.
'Provisional Atlas of Rapid Variations': 1957, from 'IAGA Symposium on Rapid Magnetic Variation in Annals of the International Geophysical Year', Vol. IIB, Pergamon Press, London.
Rastogi, R. G., Trivedi, N. B., and Kaushika, N. D.: 1966, *J. Atmospheric Terrest. Phys.* **28**, 131.
Razdan, H., Colburn, D. S., and Sonett, C. P.: 1965, *Planetary Space Sci.* **13**, 1111.
Sastri, N. S. and Jagakar, R. W.: 1967, *J. Atmospheric Terrest. Phys.* **29**, 1165.
Sato, T.: 1961, *Rep. Ionosphere Space Res. Japan* **15**, 215.
Schubert, G. and Cummings, W. D.: 1967, *J. Geophys. Res.* **72**, 5275.
Schubert, G. and Cummings, W. D.: 1969, *J. Geophys. Res.* **74**, 897.
Simon, M. and Axford, W. I.: 1966, *Planetary Space Sci.* **14**, 901.
Siscoe, G. L.: 1966, *Planetary Space Sci.* **14**, 947.
Siscoe, G. L.: 1970, *Solar Phys.* **13**, 490.
Siscoe, G. L., Davis, Jr., L., Coleman, Jr., P. J., Smith, E. J., and Jones, D. E.: 1968a, *J. Geophys. Res.* **73**, 61.
Siscoe, G. L., Formisano, V., and Lazarus, A. J.: 1968b, *J. Geophys. Res.* **73**, 4869.
Sonett, C. P. and Colburn, D. S.: 1965, *Planetary Space Sci.* **13**, 675.
Sonett, C. P., Colburn, D. S., Davis, L., Jr., Smith, E. J. and Coleman, P. J., Jr.: 1964, *Phys. Rev. Letters* **13**, 153.
Spreiter, J. R. and Alksne, A. Y.: 1969, *Rev. Geophys.* **7**, 11.
Spreiter, J. R. and Summers, A. L.: 1965, *J. Atmospheric Terrest. Phys.* **27**, 359.
Stegelmann, E. J. and von Kenschitzki, C. H.: 1964, *J. Geophys. Res.* **69**, 139.
Sturrock, P. A. and Spreiter, J. R.: 1965, *J. Geophys. Res.* **70**, 5345.
Srinivasmorthy, B.: 1960, *Indian J. Meteorol. Geophys.* **11**, 64.
Sugiura, M.: 1953, *J. Geophys. Res.* **58**, 588.
Sugiura, M., Davis, T. N., and Heppner, J. P.: 1963, paper presented at XIII General Assembly of IUGG, Berkeley, California, August 19–31.
Sura, M.: 1965, *J. Geophys. Res.* **70**, 4151.
Taylor, H. E.: 1968, *Solar Phys.* **6**, 320.
Wilcox, J. M. and Ness, N. F.: 1965, *J. Geophys. Res.* **70**, 5793.
Williams, V. L.: 1960, *J. Geophys. Res.* **65**, 85.
Willis, D. M.: 1964, *J. Atmospheric Terrest. Phys.* **26**, 581.
Yoshida, S. and Akasofu, S. I.: 1966, *Planetary Space Sci.* **14**, 979.

INTERACTION OF THE SOLAR WIND WITH THE MOON

NORMAN F. NESS*

Consiglio Nazionale delle Ricerche, Università di Roma, Laboratorio delle Plasme Spaziale

Abstract. During its orbit about the Earth, the Moon is located in the interplanetary medium or in the geomagnetosheath-geomagnetotail formed by the solar wind interaction with Earth. In the tail no evidence is found for a lunar magnetic field, which limits its magnetic moment to 10^{20} Gcm^3. ($\sim 10^{-6}$ Earth). In the interplanetary medium, no evidence exists for a bow shock or a trailing shock, although a well defined plasma wake region is observed in the anti-solar wind direction (i.e., downstream). The Moon absorbs the solar wind plasma that strikes its surface and creates a void region or cavity in the flow. Small perturbations of the interplanetary magnetic field magnitude ($< 30\%$) and direction ($< 20°$) are observed to be correlated with the location of the solar wind plasma umbra and penumbra. Characteristic perturbations in magnitude alternate in sign ($+ - + - +$) as a satellite traverses the wake region. The magnitude of the anomalies is correlated principally with the diamagnetic properties of the solar wind, as measured by β, and less with the direction of the interplanetary magnetic field. The observed lunar Mach cone gives evidence for the anisotropic propagation of waves in the magnetized collisionless warm plasma of the solar wind.

Neither the Gold-Tozer-Wilson mechanism of accretion of field lines or the Sonett-Colburn-Hollweg mechanism of unipolar induction is significant in the interaction. The transmission of microstructural discontinuities in the interplanetary medium past the Moon show little distortion, indicating a low effective electrical conductivity ($\leqslant 10^{-4}$ mho/m) which implies a relatively cool interior ($\sim 10^3$ K) of the lunar body. Fluctuations of the interplanetary magnetic field upstream from the plasma wake are stimulated by the disturbed conditions in that region. The Moon behaves like a cold, non-magnetic, fully absorbing dielectric sphere in the solar wind flow.

1. Introduction

The interaction of the solar wind with the Moon is a vastly different phenomenon from that of its interaction with the Earth (or the planet Venus). This difference is principally due to the lack of a sufficiently strong intrinsic lunar magnetic field (or ionosphere) to deflect the solar plasma flow. No bow shock wave or sheath layer of thermalized or shocked solar plasma exists around the Moon. The surface of the Moon absorbs and neutralizes the charged particle and plasma flux impacting its surface and leads to the formation of a plasma cavity and wake region trailing behind the Moon for several lunar radii. The interplanetary magnetic field is at times slightly perturbed in magnitude ($< 30\%$) but little in direction within and near the solar wind wake ($< 20°$); the effects depend entirely upon the characteristics of the magnetized, warm anisotropic solar plasma. The Moon behaves much like a non-magnetic, electrically non-conducting, dielectric, fully absorbent, spherical obstacle in the solar wind flow.

This paper presents a review of the current state of experimental and theoretical studies of the interaction of the solar wind with the Moon. Essentially, data from one lunar orbiting satellite, the U.S.A. Explorer 35, placed in orbit in July, 1967 have provided all the results and interpretations upon which this understanding is based. In

* At the time of the Symposium the author was on leave of absence from the NASA Goddard Space Flight Center, Greenbelt, Maryland, U.S.A., to which he has since returned.

order to place contemporary views in the proper context, a review of earlier satellite studies directly related to the subject of this paper is also included. The reader interested in the long studied problem of possible lunar effects in terrestrial magnetic field variations is referred to the recent review by Schneider (1967). There, the numerous studies of the correlation between lunar phase and the planetary magnetic activity indices K_p and C_i are summarized. Those statistical research efforts have not contributed significantly to our present understanding, because of the weakness of the interaction that in fact occurs between the solar wind and the Moon, and the small size of the Moon when compared to the Earth's magnetosphere and tail. Attempts to simulate the problem in the laboratory are discussed by Fahleson (1967).

Observations of the solar wind interaction with the Moon are possible only during that part of the lunar synodic month when the Moon is outside the geomagnetotail and magnetosheath. This normally occurs for about 10 days before and after new Moon, depending upon the motion and position of the Earth's bow shock. It is possible to observe the interaction of the thermalized solar plasma with the Moon for an additional 6 days when the Moon is located inside the Earth's magnetosheath. For the remaining 3.5 days, the Moon is located in the geomagnetotail and occasionally imbedded in the plasma sheet associated with the field-reversal region of the tail. When the Moon is in the tail, it is shielded from the solar wind and it is possible to measure most effectively the magnetostatic properties of the lunar body. That there are four characteristic environments of magnetized plasma for the Moon has been recognized only since 1964, following the initial studies of the solar wind interaction with the Earth.

In the absence of a bow shock wave (like that which strong interaction generates) and of an appreciable atmosphere, the flow of the magnetized solar plasma past the Moon depends upon the electrodynamic properties of its interior and the coupling between its surface and the adjacent plasma. Preliminary attempts have been made to deduce the electrical conductivity from the presence or absence of certain features in the characteristics of the solar wind flow. The results suggest a moderately low conductivity of 10^{-4} mho/m or smaller, for the deeper interior and much smaller for the layers near the surface. In spite of the remarkable increase in our understanding of the topic of this review in the past three years, there are unresolved problems in the interpretations of the experimental data.

This paper begins with an historical review of early studies in the satellite era, from 1959–1966. Next a discussion of the experimental findings by Lunar Explorer 35 are presented with their interpretations. Then the present status of studies of the electrical conductivity of the interior of the Moon is discussed. The paper concludes with a summary of our present understanding and problems for future study.

2. Early Studies: 1959–1967

The first direct observational measurements related to the solar wind interaction with the Moon were those performed by the U.S.S.R. lunar impacting probe Luna 2 in

1959. Subsequently in 1963–1964 the Earth-orbiting IMP-1 (Explorer 18) satellite reported possible observations of the distant lunar wake. The first spacecraft placed into lunar orbit, the U.S.S.R. Luna 10, provided intermittent measurements of the magnetic field and low-energy plasma for a period of two months in 1966. None of these experimental studies was either sufficiently sensitive or accurate to reveal the true nature of the solar wind interaction with the Moon. During the period 1964–1967, several theoretical studies of this problem were made. While covering a wide range of possibilities, most of them were not correct in their anticipation of the characteristics of the interaction. This section will briefly summarize these early satellite studies and also the early theoretical models proposed.

2.1. LUNA 2 1959

The U.S.S.R. launched two probes in 1959 which were designed to study the Moon: Luna 1, on 2 January and Luna 2 on 12 September. Both spacecraft carried a three-component magnetometer but Moon-related results were reported only from Luna 2, which impacted on the surface on 13 September, 1959. (Luna 1 achieved a fly-by of the Moon on 3 January at distance of 7000 km.) According to the Luna 2 magnetic field studies, the Moon possessed a magnetic field no larger than 100γ on the basis of measurements made up to 50 km from its surface (Dolginov *et al.*, 1961). The errors associated with the instrumentation and the spacecraft magnetic field limited the accuracy to approximately 50–100 γ (Dolginov *et al.*, 1960). While this result indicated that the Moon possessed a much weaker intrinsic magnetic field than the Earth, it was pointed out by Neugebauer (1960) that if the solar wind compressed the lunar field then the field might in fact be much stronger.

At the time these measurements were performed, the general problem of the solar wind interaction with the Earth was not understood and so the advantage of a comparative analysis could not be enjoyed. We now know that a lunar surface field of 50 γ is more than sufficient to deflect the solar wind and lead to the formation of a detached bow shock wave. But in 1959, these problems were only beginning to be defined. The absence of a lunar field stronger than the limit set by Luna 2 was not surprising because of the very low rotation rate of the Moon and its low average density. This precludes an appreciable core and a dynamo mechanism similar to the Earth's that might develop and generate a lunar field.

The magnitude of the transverse magnetic field B_T that is sufficient to deflect the solar wind flow is determined by equating the directed pressure of the solar wind stream of particle density n and velocity V_{sw} to that of the magnetic field

$$\frac{(2B_T)^2}{8\pi} = knm_p V_{sw}^2 \cos^2 \phi, \tag{1}$$

where m_p represents the mass of the proton, the principal constituent of the solar wind. The factor of 2 in the left hand side of Equation (1) is due to the induced electrical currents which arise as the magnetic field separately deflects the ions and electrons in opposite directions. The factor k is theoretically 2 for perfect reflection of the plasma from the boundary surface (which is assumed to be oriented with its normal at an angle ϕ to the solar wind). Our present understanding of the solar wind interaction with the Earth indicates that $k=0.88$ is a better value to use (Spreiter and Alksne, 1969), because it indicates deflection around an obstacle rather than reflection from it. With our present knowledge of solar wind properties, an assumed density of $10/\text{cm}^3$ and a velocity of 400 km/s implies a B_T of approximately 30γ for $\phi=0°$, i.e. normal incidence.

Neugebauer (1960) used higher estimates of $n=1000/\text{cm}^3$ and $V_{sw}=500$ km/s, consistent with views of the solar corpuscular flux held at that time. A more significant aspect, however, was that the existence of the Earth's bow shock wave had been neither predicted nor measured in 1959–1960. Thus the most important feature of the Luna 2 data, the absence of a lunar bow shock, was not considered in the interpretation of the results. Had it been, then the upper limit of the intrinsic lunar field could have been set much lower.

A final point to consider in this review of the Luna 2 measurements is that of the phase or position of the Moon at the time the measurements were made. The Sun-Earth-Moon angle was 140° East and thus the Moon was probably located in the Earth's magnetosheath rather than the interplanetary medium. However, the errors in the Luna 2 measurements were so large that further consideration of this uncertainty is not meaningful.

2.2. IMP-1 (EXPLORER 18)

From November, 1963 to February, 1964, the U.S.A. spacecraft IMP-1 periodically measured the interplanetary medium from its highly eccentric Earth orbit. The synchronism and geometry of its orbit were such that from December, 1963 to February, 1964 apogee was located near the extension of the Sun-Moon line. Ness *et al.* (1964) suggested that an observed interval of enhanced and disturbed interplanetary magnetic field during the period 13–15 December might have been due to the passage of IMP-1 through the lunar wake since the satellite was favorably situated at a distance of 150 R_M (R_M=lunar radius=1738 km) to lie within the solar wind interaction region. A subsequent study of the data during the January and February opportunities did not yield substantiating evidence for a similar disturbance (Ness, 1965a). In January, however, the satellite was not favorably situated and in February it was much more distant from the Moon (at 200 R_M).

The interpretation of a lunar wake thus suggested the existence of a lunar bow shock wave and an intrinsic magnetic field. Several authors questioned the interpretation of the observed disturbance as being due to the solar wind interaction with the Moon. Greenstadt (1965) suggested that a small solar flare, which occurred before the IMP-1 observations, might have been responsible for the disturbed conditions. Hirshberg (1966) offered the alternative explanation that the disturbance was one of a continuing series of *M*-Region disturbances. Ivanov (1965) postulated that the disturbance might even have been due to the solar wind interaction with the Earth and associated with disturbances propagating upstream from the Earth. Ness (1965b, 1966a and

1966b) replied to these comments and, in the case of Ivanov's suggestion, showed that the flow of solar wind was sufficiently super-Alfvénic to eliminate the Earth's interaction as the source. Michel (1965) examined the positions of the Moon and IMP-1 and found that the conclusions of Ness were not inconsistent with typical solar wind parameters.

With our present knowledge of the solar wind interaction with the Moon, it appears certain that the lunar association ascribed to the IMP-1 data was incorrect and that the disturbance was indigenous to the solar wind. (Michel, 1968b; Taylor *et al.*, 1968). Whether the source was a solar flare or an *M*-Region disturbance has not been determined.

2.3. THEORETICAL STUDIES: 1964–1967

As the continual flow of the magnetized solar wind came to be recognized as a permanent phenomenon in interplanetary space, studies of its interaction with the Moon and its atmosphere were begun (see Figure 1). Michel (1964) first suggested that the solar wind flow past the Moon would have characteristics somewhere between the following limits:

(1) *Undeviated flow*, with the solar wind being completely absorbed and neutralized upon impacting the lunar surface, as seen in Figure 1a;

(2) *Potential flow*, in which the plasma flow would be deflected around the Moon by a magnetic field induced as the interplanetary magnetic field is dragged past a lunar interior having finite conductivity. The development of a bow shock would lead to subsonic flow behind it, as shown in Figure 1b.

The principal effect of the interplanetary magnetic field was considered to be the increased cross-sectional area that would result from the finite Larmor radii for the solar wind ions. Since the major topic of Michel's study was the lunar atmosphere, and because so little was known at that time about the solar wind and its interactions, he did not further develop these two limiting models for the interaction. In both cases, the moon absorbed the solar plasma which impacted its surface.

In another study of the interaction Gold (1966) suggested that because of the high internal electrical conductivity of the Moon, the interplanetary magnetic field lines would be caught up in the lunar interior. This would result if the diffusion times of the magnetic field (Cowling's $\tau_D = \mu \sigma L^2$, in which L is a characteristic scale-length for the Moon) were very long compared to the time for convection of the solar wind plasma past the Moon ($\tau_c = 2R_M/V_{sw}$). Gold (1966) postulated the existence of a detached bow shock wave and a lunar magnetosheath, since he assumed that the conductivity was sufficiently high to cause accretion of the interplanetary magnetic field.

The mechanism of Gold (1966) for the formation of a lunar bow shock, a pseudo-magnetosphere, and a tail was re-examined by Tozer and Wilson (1967). They noted that the electrical conductivity estimates for the Moon which Gold had used were unreasonably high. Using lower estimates based on a comparison with the Earth's mantle, they concluded that at most a small internal core would be conductive enough to entrap the interplanetary magnetic field, mainly because of the higher temperatures in the core. As a result, no detached bow shock would develop, but only an attached shock tangent to the surface of the Moon, as shown in Figure 1c. They still envisaged the development of an accreted magnetic field and a pseudo-magnetosphere.

Present understanding of the solar wind interaction indicates that of these early theoretical models, the one most closely approximating the actual flow characteristics is that of undeviated flow, one of Michel's two 1964 alternatives.

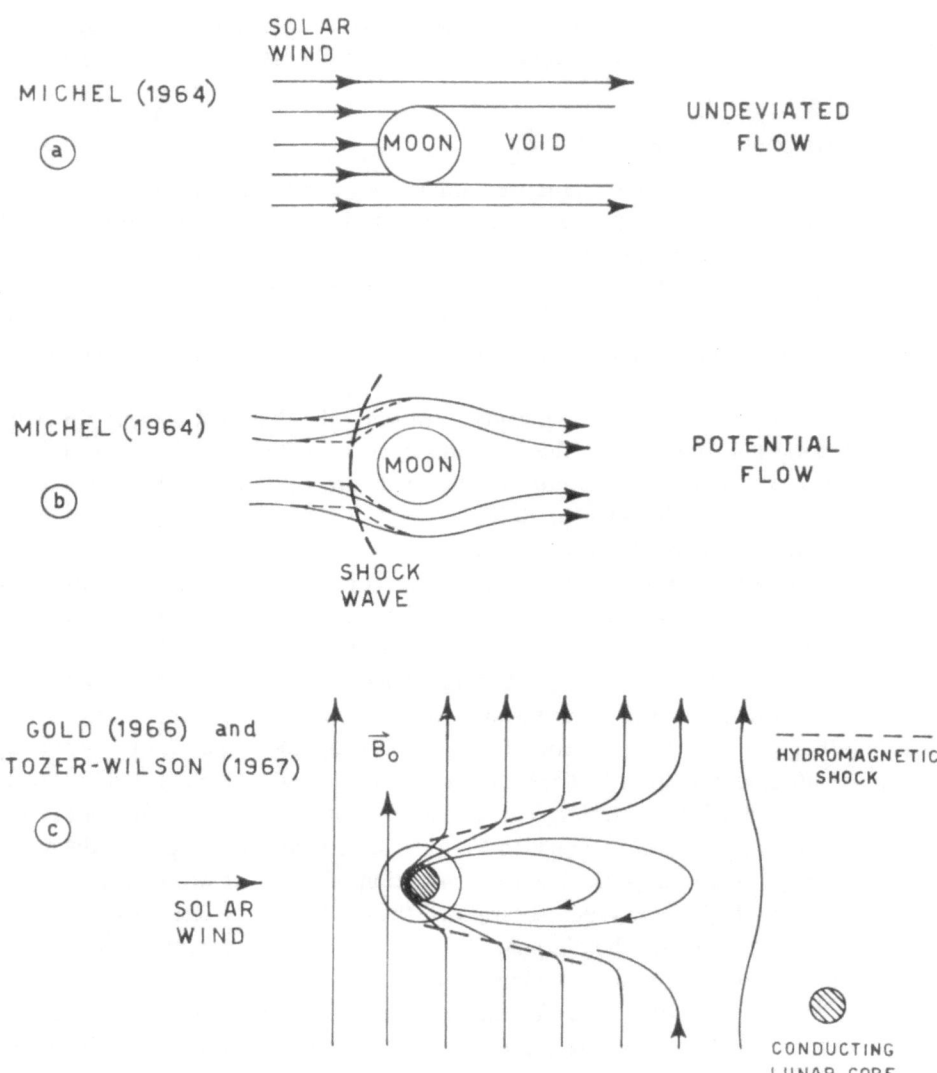

Fig. 1.   Some theoretical suggestions for structure of solar wind interaction with Moon. The predicted effect of the interplanetary magnetic field in models (a) and (b) was to increase the effective size of the Moon with respect to flow of the solar wind around the obstacle. The dashed lines in (b) indicate the effect of the detached bow shock wave.

2.4. Luna 10

On 2 April, 1966, the U.S.S.R. placed the Luna 10 spacecraft into a close lunar orbit with periselene (point of closest approach) = 350 km, aposelene = 1017 km and period = 3 hr. The data transmission of magnetic field, plasma, and energetic particle measurements was intermittent because battery life was limited and there was no solar array, but the measurements were distributed over a time interval of 36 days. The results from the plasma detector were interpreted as indicating the presence of the geomagnetic tail for measurements taken near full Moon (Gringauz et al., 1966). But there were no reports of a variation with Sun-Moon-Probe angle of the detected fluxes, regardless of the phase of the Moon. The decrease in the low-energy electron flux near full Moon when compared to measurements in interplanetary space was interpreted to mean the presence of the geomagnetic tail.

The results from the magnetic field experiment (Zhuzgov et al., 1966) were interpreted in terms of a permanent lunar pseudo-magnetosphere which was carried along with the Moon throughout its entire orbit. In spite of uncertainties in the measurements of 10γ, the results are believed to indicate the presence of relatively constant fields parallel and perpendicular to the spin axis of the spacecraft several times larger than this, and also therefore much larger than the typical interplanetary field of 6γ. Since no on-board aspect system provided directional data, a study of the orientation of the observed field was not possible. These large, steady fields were interpreted on the basis of Gold's (1966) suggestion of a trapped interplanetary magnetic field.

Ness (1967) raised a question regarding the interpretation of a pseudo-magnetosphere and the failure of the magnetic field experiment to observe the geomagnetic tail, since simultaneous measurements by IMP-3 (Explorer 28) in April and May 1966 showed conclusively that the tail extended essentially uniformly from the Earth to 38 R_E, two-thirds of the distance to the Moon ($\sim 60 R_E$).

From a study of the position of the Moon in the geomagnetic tail region, Ness (1967) reached the plausible assumption that Luna 10 was located in the plasma sheet of the tail at the times of measurements. Hence it was not sensitive to the magnetic field of the tail. Dolginov et al. (1967) reviewed their earlier interpretations and agreed with this viewpoint, but still maintained that a lunar pseudo-magnetosphere existed. It was shortly afterwards that the U.S.A. launched Explorer 35 successfully into lunar orbit, and those results will be discussed in the next section.

3. Recent Studies: 1967 to the Present

The U.S.A. spacecraft Explorer 35, placed into lunar orbit on 22 July, 1967, provided the first accurate measurements of the nature of the solar wind interaction with the Moon. The 104 kg magnetically clean spin-stabilized spacecraft was the second in a special set of two modified satellites of the IMP (Interplanetary Monitoring Platform) series. Known as IMP-D and E before launch, they included a fourth stage solid propellant retromotor for injection into an orbit around the Moon. The first attempt

on 1 July, 1966 failed and Explorer 33 achieved instead a highly eccentric earth orbit
with apogee greater than 80 R_E ($1 R_E = 6378$ km).

The second launch, on 19 July, 1967, was successful and Explorer 35 was injected
into a very stable lunar orbit with period 11.5 hr, inclination $= 169°$, aposelene $= 9388$
± 100 km (5.4 R_M) and periselene $= 2568 \pm 100$ km (1.4 R_M). The principal perturbation consists of a fortnightly modulation of the semimajor axis of the orbit due to the
gravitational masses of the Earth and Sun, essentially a tidal effect. As the Earth-
Moon system orbits the Sun, the aposelene-Moon-Sun angle (direction of the semimajor axis relative to the Sun) changes annually by $> 360°$ so that the entire region of
solar wind interaction from periselene to aposelene is sampled throughout the year.
The spacecraft carried magnetometers, energetic particle detectors, and plasma probes
which have functioned continuously during the more than 2100 orbits since launch, and
data have been received continuously except for periods of occultation behind the
Moon when reception is blocked. The spin axis of the spacecraft is normal to the
ecliptic ($\pm 2°$) and the spin rate is 25.6 ± 0.4 rev/min. An earlier review (Ness, 1969)
briefly summarized the scientific results from all of the experiments on Explorer 35 up
through May 1968. The present paper extends the review of Explorer 35 data through
April 1970, although from only the magnetometers, plasma probe, and energetic
particle detectors. In addition a review of relevant theoretical studies is included.

3.1. EARLY MAGNETIC FIELD-PLASMA RESULTS: EXPLORER 35

Preliminary reports on the NASA-GSFC magnetic field experiment (Ness *et al.*, 1967)
and the MIT plasma probe (Lyon *et al.*, 1967) were presented at the NASA Santa
Cruz Summer Study on Lunar Exploration held in August-September, 1967. These
results showed

(1) The *absence of a lunar magnetic field* (at least none greater than 2γ) at satellite
periselene when the Moon was in the geomagnetic tail;

(2) The *absence of a bow shock wave* or magnetosheath (similar to the Earth's)
surrounding the Moon, when it was in the interplanetary medium (see Figure 2a) and

(3) The *existence of a plasma cavity* or void region behind the Moon when in the
solar wind flow (see Figure 3).

The first results indicated that the Moon did not possess a dipole moment greater
than 4×10^{20} cgs units ($M_E = 8 \times 10^{25}$ cgs) so that no deflection of the solar wind
would occur when the Moon was in the geomagnetosheath or solar wind flow (see
also Sonett *et al.*, 1967). No significant disturbance of the magnitude of the magnetic
field was noted as the spacecraft passed through the solar wind flow downstream
from the Moon. The only effect noted in the interplanetary field was the existence,
sometimes, of:

(4) *Field magnitude increase in the region corresponding to the plasma umbra* and

(5) *Field magnitude decreases on either side*, in the plasma penumbra.

These anomalies were interpreted in terms of the diamagnetic properties of the
solar plasma (see also Colburn *et al.*, 1967).

The absence of a pseudo-magnetosphere was interpreted by Ness *et al.* (1967) to mean that the effective electrical conductivity of the Moon was quite low, less than 10^{-5} mho/m, and that the Moon did not entrap the interplanetary magnetic field, as suggested earlier by Gold (1966) and by Tozer and Wilson (1967). The results also indicated that there existed no detached bow shock as suggested in one of Michel's

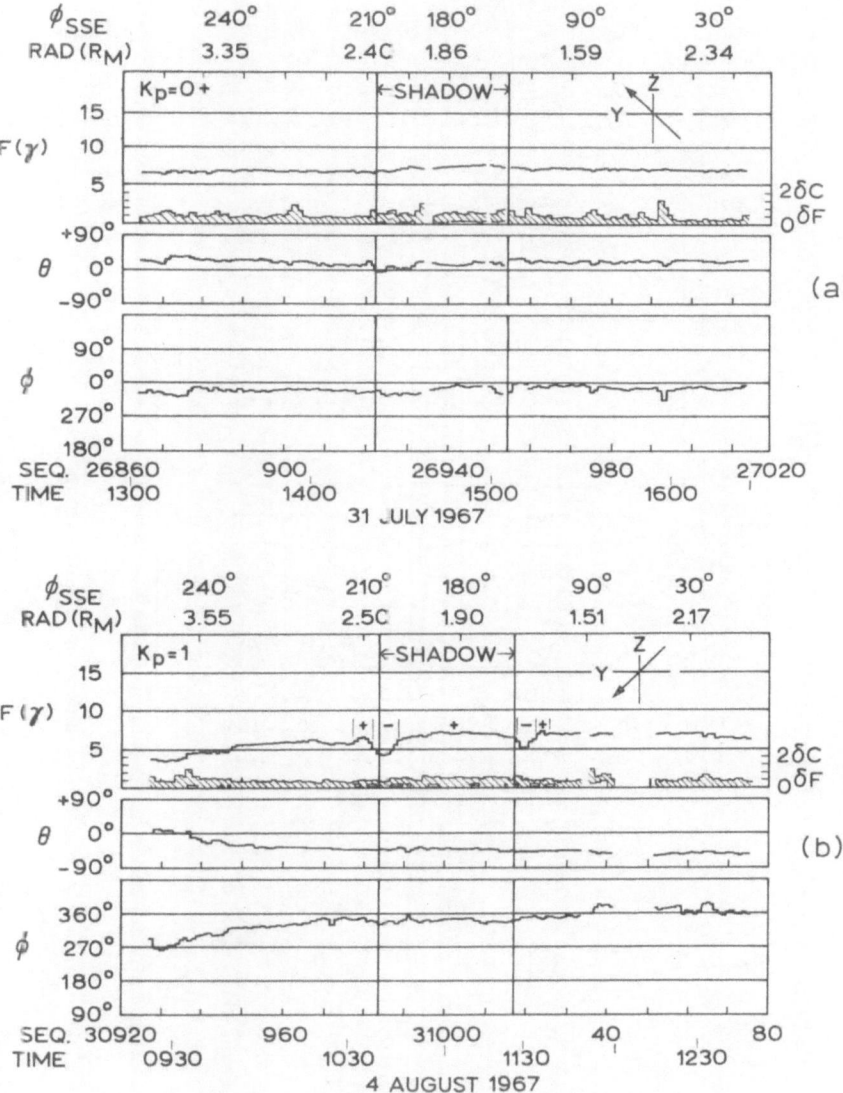

Fig. 2.   Early Explorer 35 studies of the magnetic field in the vicinity of the Moon as the solar wind flows past. There is no evidence for a bow shock; the direction (θ,ϕ) and magnitude (F) of the field are only slightly perturbed in the leeward region of solar wind flow (Ness et al., 1968). The selenocentric longitude (ϕ_{SSE}), and radius (in lunar radii R_M) of the spacecraft are given on the scales at top, and times on the bottom scales.

1964 models (see Figure 1b) nor a shock wave tangent to the lunar surface as suggested by Tozer and Wilson (see Figure 1c)

The determination of the topology of the magnetic field in what is now called the *Lunar Wake* region was complicated by the spacecraft spin rate behavior when located

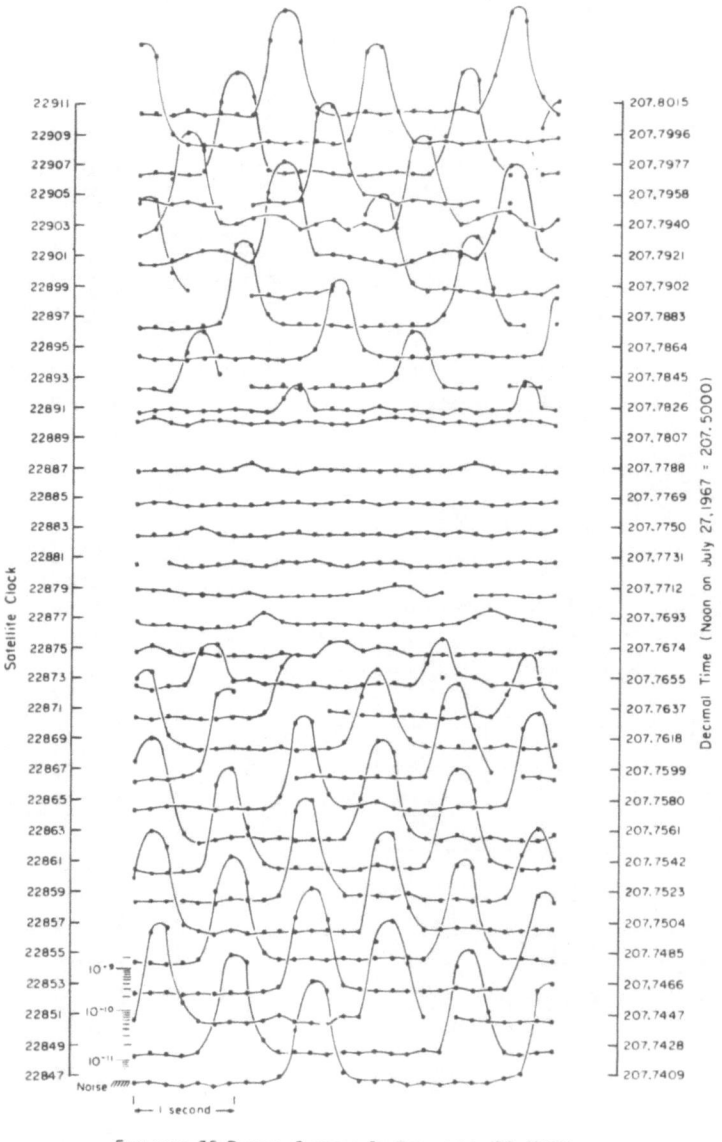

EXPLORER 35 PLASMA CURRENT SAMPLES (July 27, 1967)

Fig. 3.   Early Explorer 35 studies of the integral solar wind plasma flux ($50 < E_p < 2850$ eV) on the downwind side of the Moon (Lyon *et al.*, 1967). A logarithmic scale, used for the amplitude of the plasma current, is shown at lower left.

in the optical shadow of the Moon. Although the spacecraft provided a pseudo-Sun reference signal when in shadow, based upon the last measured spin period when illuminated, the spin rate of the spacecraft changed during the shadow crossing. This was due to the physical contraction of the spacecraft after removal of the solar thermal input, the associated decrease in angular moment of inertia and the concomitant

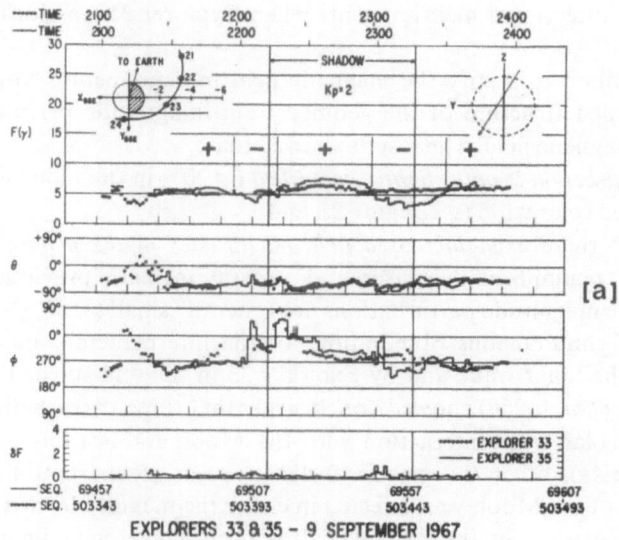

EXPLORERS 33 & 35 – 9 SEPTEMBER 1967

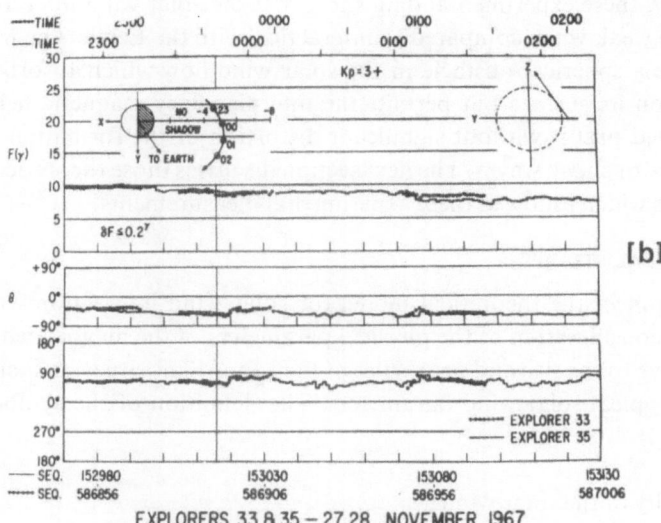

EXPLORERS 33 & 35 – 27,28 NOVEMBER 1967

Fig. 4. Simultaneous measurements of the interplanetary magnetic field by Explorers 33 and 35 as the latter spacecraft passes through the lunar wake (Taylor *et al.*, 1968). Dots indicate Explorer 33 data and time scale (above each plot), and solid lines those for Explorer 25.

increase in spin rate to conserve angular momentum. When the spacecraft reentered sunlight, the spin rate decreased to its normal pre-shadow value. The change in spin period was only 0.1% of the normal value but since this was integrated through the shadow passes (up to 1 hr in length), the net result was a linearly increasing angular error in the azimuthal angle ϕ_{SE} of up to more than 360° at the end of the shadow. Taylor (1968) developed an algorithm to determine the corrections to be made to the directional measurements when Explorer 35 was located within the shadow.

Using this method to rectify the magnetic field measurements, Ness *et al.* (1968) studied the detailed structure of the geometry and magnitude perturbations of the interplanetary magnetic field. The results showed that:

(6) The *field direction is only slightly perturbed* (<20°) in the lunar wake from that in the undisturbed solar wind (see Figure 2), and

(7) Sometimes there exist *increased field magnitudes in the penumbral regions* in addition to the penumbral decreases and umbral increase previously noted (see Figure 2b). These magnitude perturbations are generally small (<30%).

In a study of simultaneous observations of the interplanetary magnetic field by Explorer 35 in the lunar wake and by Explorer 33 in the undisturbed interplanetary medium, Taylor *et al.* (1968) showed conclusively that these perturbations were associated with the solar wind interaction with the Moon and not intrinsic to the solar wind (see Figure 4a). When the spacecraft distance was greater that 4 R_M behind the Moon, no effect of the Moon was detected in either the magnitude, direction, or rapid fluctuations as measured by the RMS deviation over an averaging interval of 81.8 sec (see Figure 4b).

In summary, these experimental data show that the solar wind interaction with the Moon is quite weak when compared to interaction with the Earth. The Moon appears to behave like a spherical obstacle in the solar wind flow which absorbes the plasma flux incident on its surface but permits the interplanetary magnetic field to be convectively carried past it without significant distortion or the formation of a pseudo-magnetosphere or shock waves. The next section discusses those theoretical studies and considerations which followed these experimental measurements.

3.2. THEORETICAL STUDIES

An investigation of the theoretical model for solar wind interaction with the Moon begins with a consideration of the physical parameters of the magnetized solar plasma at 1 AU relative to the size and properties of the Moon. The situation is summarized in Figure 5 for typical solar wind parameters. The definition of the symbols used is as follows

V_{sw} – Velocity of the solar wind;
n – Number density of ions (electrical neutrality assumed);
B_0 – Interplanetary magnetic field at angle ϕ_0 to Moon-Sun line;
T – Temperature (parallel, $T_\parallel$, and perpendicular, $T_\perp$, to the magnetic field);

V_T – Thermal velocity of particles (root-mean-square value) corresponding to T;
R_L – Larmor (cyclotron) radius;
τ_L – Larmor (cyclotron) period;
L – Characteristic scale-length of wake;
τ – Characteristic time scale, required to convect past Moon;
β – Ratio of plasma pressure (perpendicular to field line) to perpendicular magnetic field pressure;
V_A – Alfvén velocity.

It is seen that the Larmor radius is small in comparison with the radius of the Moon ($<2\%$). Thus, from a kinetic plasma viewpoint, a guiding center approximation would be valid in a first approximaticn since it is also known from experimental observations that the distances over which the small changes in field magnitude and direction occur is large compared to the Larmor radius.

From a continuum approach, a measure of the *effective* mean free path for particle interactions determines the scale length for collective fluid-like phenomena. This may range from the Debye length (~ 1–10 m) to the Coulomb collision mean free path ($\sim 10^7$ km). The correct value is not known exactly, but the success of fluid descriptions of solar wind flow past the Earth suggest that it is very much less than the size of the magnetosphere. The existence cf classical magneto-hydrodynamic disconti-nuities in the interplanetary medium has been interpreted in terms of an interaction

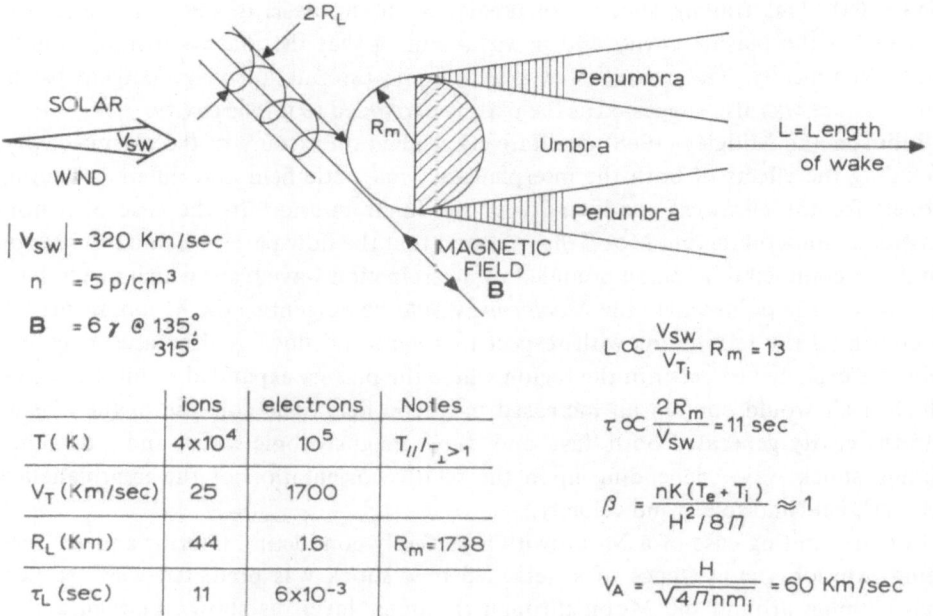

Fig. 5.   Summary of physical considerations for theoretical analysis of solar wind flow past the Moon, using the most probable values of plasma and magnetic field parameters.

scale length less than the 'thickness' of the thinnest of these, 500 km (Siscoe *et al.*, 1969). Many theoretical studies have assumed that a continuum-fluid description is appropriate in the analysis of the solar wind interaction with the Moon.

In either approach, there is agreement that:

(1) The surface of the Moon is modeled as a perfect (or nearly perfect) absorber of particle or plasma fluxes and

(2) The flow of the solar wind is highly supersonic with respect to the velocity of propagation of disturbances in the medium.

Thus, item (1) assures that there can be no bow shock wave and item (2) that the region of disturbed solar wind flow will lie principally behind the Moon centered on its geometric shadow with respect to the solar wind flow.

Sonett and Colburn (1967) studied the induction of an electrical current flow and the formation of an associated bow shock wave resulting from the convective transport of the interplanetary magnetic field past the Moon. They showed that the interaction with the Moon could range from weak to strong, with the formation of a magnetosheath and the development of a shock wave for values of a homogeneous interior electrical conductivity between 10^{-6}–10^{-3} mho/m. No specific model for the Moon was proposed, although it was noted that a unipolar mechanism could not function for planets with insulating mantles.

Shortly after the early results from Explorer 35 were reported (see Subsection 3.1 above), Michel (1967) proposed that as the solar wind closed in behind the Moon, a shock wave would be generated when the collapse of the plasma was halted (see Figure 6a). This trailing shock was predicted on the basis of zero magnetic field pressure in the plasma cavity and it was assumed that the plasma behaved purely hydrodynamically. The existence of a hypersonic rarefraction wave tangent to the lunar surface was also suggested as the plasma expanded to fill the cavity.

Johnson and Midgley (1968) qualitatively studied the closure of the plasma cavity, including the effects of both the interplanetary magnetic field and different limiting models for the electrical conductivity of the lunar interior. In the case of a non-magnetic, non-conducting Moon they showed that the flow pattern behind the Moon would be confined to a region bounded by a rarefaction wave front which would have its origin at the point where the *Mach cone* would be tangent to the Moon, instead of at or behind the terminator with respect to solar wind flow. A decreased magnetic field was expected to occur in the region where the plasma expanded to fill the cavity, which itself would contain an increased magnetic field. The collapse of the plasma into the cavity generated both 'fast' and 'slow' magnetosonic waves and a detached trailing shock wave, depending upon the relative orientation of the interplanetary magnetic field and solar wind velocity.

For the limiting case of a Moon with a perfectly conducting interior and an insulating exterior, the existence of a detached bow shock was predicted with the field lines slipping around the Moon through the outer layer, as shown in Figure 7. A number of other models for the electrical conductivity were also considered in which the major features of the flow pattern behind the Moon were unchanged.

In an extension of his earlier study, Michel (1968a) included the effects of the magnetic field in modifying the geometry of the previously proposed trailing shock (see Figure 6b). Regardless of the orientation of the field, a trailing shock was expected as the solar plasma collapsed to fill the cavity region. In between the rarefaction wave and the cavity, the decreased density of the expanding plasma led to a decrease of the component of the magnetic field transverse to the direction of the expansion but there was no effect on the parallel component. Within the cavity an increased magnetic field was expected to result from the compression of the vacuum magnetic field by the magnetized solar plasma.

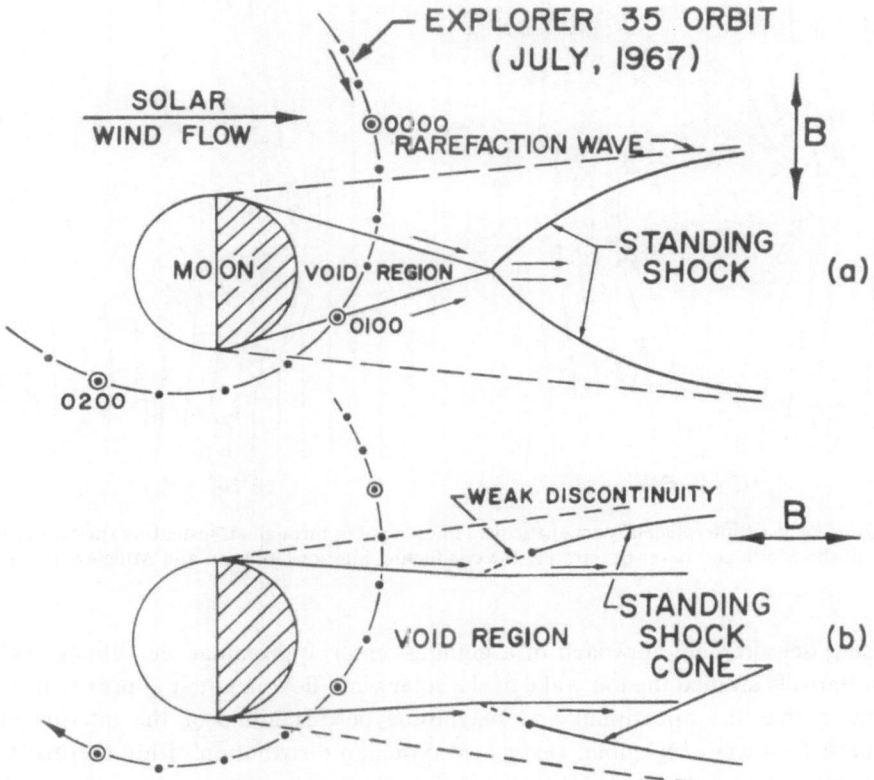

Fig. 6. Sketch of the development of a trailing shock wave in the lunar wake: (a) according to Michel (1967); (b) according to Michel (1968a).

In studying the geometry of the boundaries of the rarefaction wave and plasma void, Michel (1968a) proposed that they would be elliptical in shape (see Figure 8). This is due to the difference in the propagation velocities of the magneto-acoustic mode, V_{MA}, and the acoustic modes, V_S. We have $V_{MA} = \sqrt{V_A^2 + V_S^2}$, where $V_A = B_0/\sqrt{4\pi\varrho}$ is the Alfvén velocity and $V_S = \sqrt{\gamma p/\varrho}$ is the acoustic velocity. Here ϱ is the mass density, p is the pressure, and γ is the ratio of specific heats. The factor C in Figure 8 is different for each velocity and is related to the parameter γ and the ratio of the total pressure inside the plasma void to the pressure outside. For nominal solar wind parameters and expansion into a vacuum, $C=3$ for the acoustic mode. For expansion into a region having a pressure one-half the ambient pressure, $C=0.4$, for the magneto-acoustic mode.

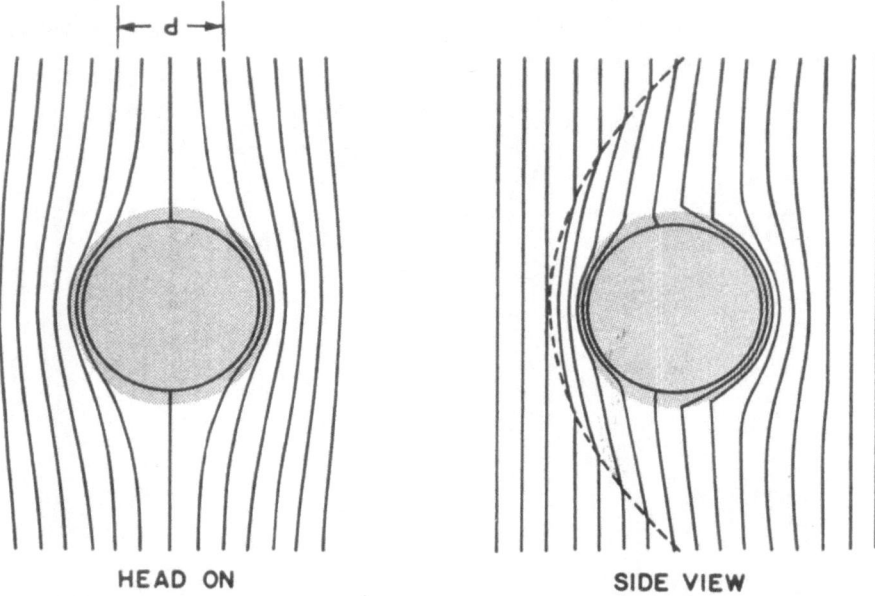

HEAD ON   SIDE VIEW

Fig. 7.   Sketch of interplanetary magnetic field lines slipping through an insulating shell outer layer of the Moon and never penetrating the conducting interior (Johnson and Midgley, 1968).

Using the alternate approach of a guiding-center approximation, Whang (1968a) quantitatively studied the ion wake of the solar wind flow in a first approximation by assuming that the directional and magnitude perturbations of the interplanetary magnetic field were negligible. Using a Maxwellian distribution of ion thermal velocities with mean value $V_{\|i}$ and defining the speed ratio $S_i = V_{sw}/V_{\|i}$, he determined the three-dimensional geometry of the plasma wake as a function of the orientation of the magnetic field, ϕ_0, relative to the solar wind velocity. These results showed that the wake was confined to a region down wind from the Moon (see Figure 9a) and was not axially symmetric. The predicted properties of the ion wake were the following:

(1) It will be elongated to greater than a lunar diameter in the plane of symmetry defined by the magnetic field direction and solar wind velocity (the normal $\mathbf{n} = \mathbf{V}_{sw} \times \mathbf{B}_0$) (see Figure 9a).

(2) It will be restricted to exactly 1 lunar diameter in thickness, normal to this symmetry plane (see Figure 9b).

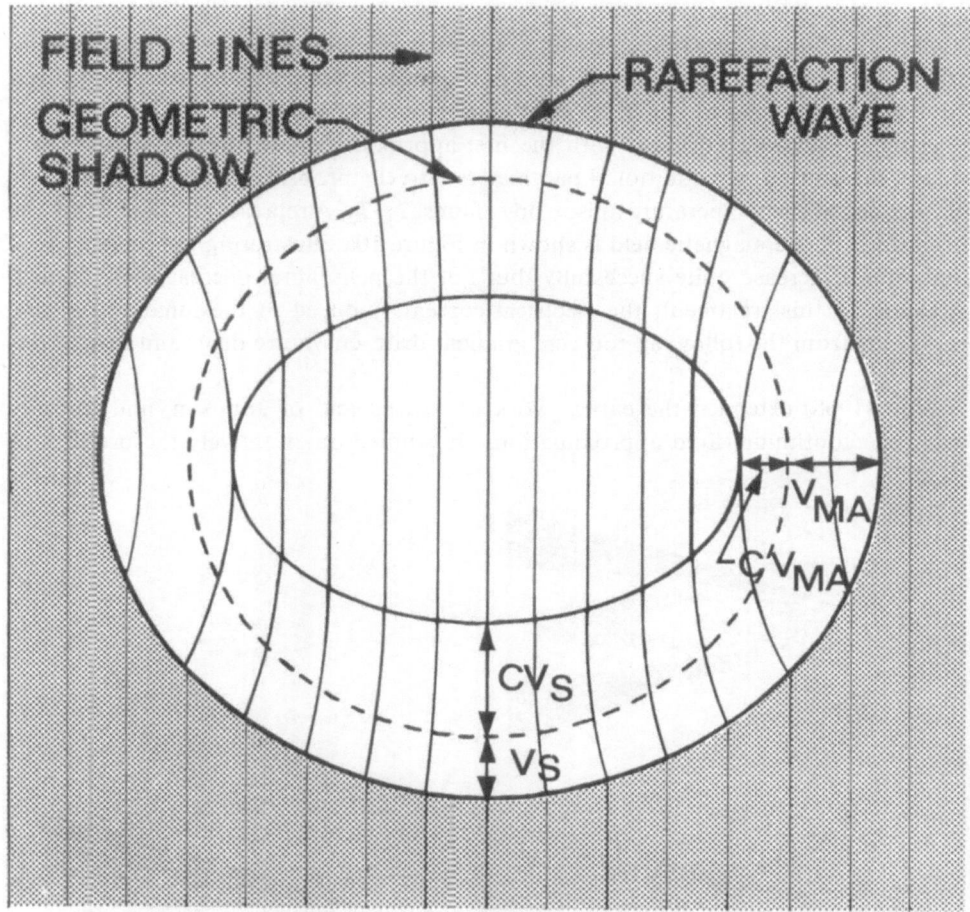

Fig. 8.   Diagram illustrating the elliptical cross-sectional geometry of lunar rarefaction and recompression waves as solar wind flow expands into cavity region (Michel, 1968a). For more detailed explanation, see text.

(3) The axis of the plasma cavity, elliptical in cross-section normal to the axis, deviates slightly from parallellism to the solar wind velocity vector.

(4) The thermal anisotropy of the plasma increases substantially in the cavity since only those particles with high thermal velocity parallel to the field can enter the cavity region.

(5) The direction of plasma flow becomes markedly deviated towards the axis of the cavity within the cavity region.

The length of the wake depends upon ϕ_0 and is longest (theoretically, infinitely long) when $\Phi_0 = 0°$ (or 180°), and shortest when $\phi_0 = 90°$ (or 270°). The possibility of plasma instabilities due to the modified distribution functions in the lunar wake was not considered.

Using this first-approximation representation of a disturbed ion wake, Whang (1968b) and Ness *et al.* (1968) computed the magnetic field perturbations that would arise from the electrical currents induced in the wake. These computations were based on a cylindrical Moon; the final field configuration, derived iteratively, satisfied Maxwell's equations cosistent with the first approximation of the ion wake. These studies introduced two additional parameters into the problem; namely: the β of the plasma and η, the temperature anisotropy of ions, $T_\parallel/T_\perp$. A representative profile of the magnitude of the magnetic field is shown in Figure 10a, illustrating the prediction of the umbral increase quite succesfully, but not the penumbral decreases of the field strength. In this treatment, the electrical currents induced in the lunar wake were computed from the following sources: gradient drift, curvature drift, and magnetization.

Wolf (1968) extended the earlier work of Michel and of Johnson, and Midgley using the continuum fluid approximation. He studied quantitatively the problem of

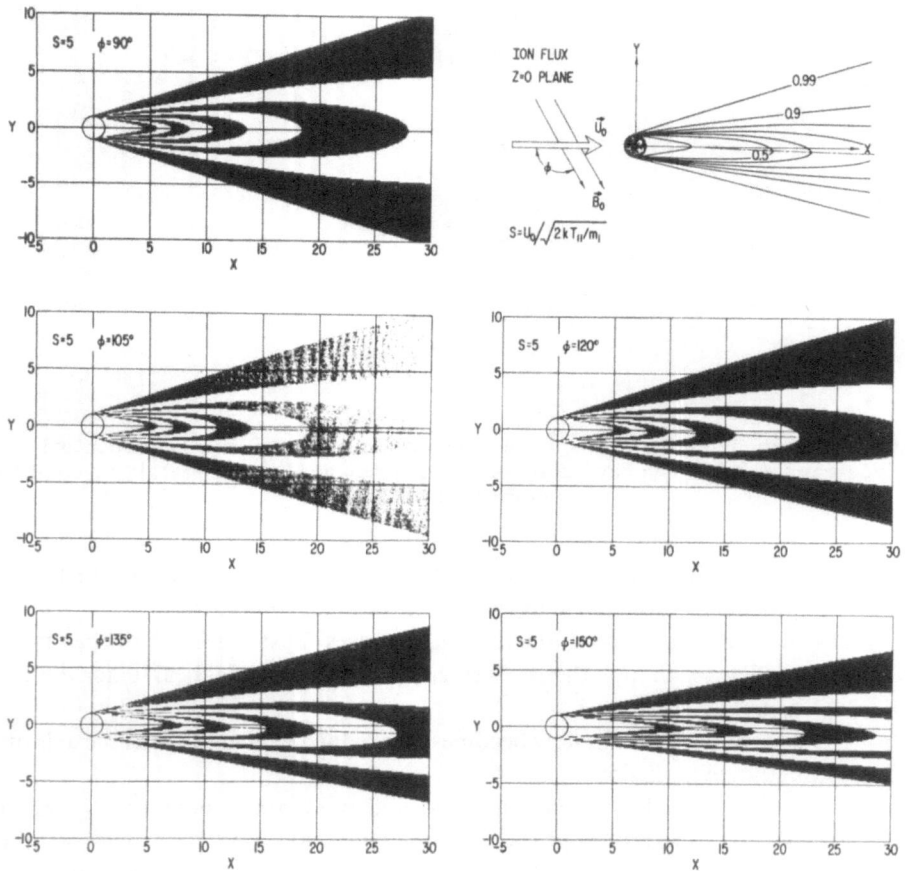

Fig. 9a.

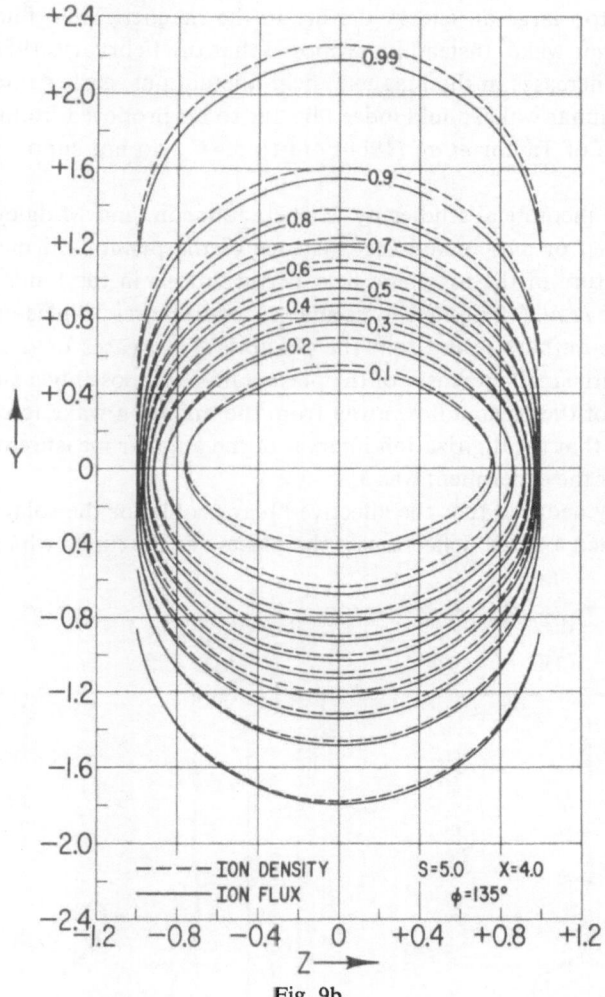

Fig. 9b.

Fig. 9a–b. Three-dimensional properties of ion wake shown: (a) in the plane of symmetry (the xy-plane) and (b) transverse to plane of symmetry (the yz-plane) (Whang, 1968a).

flow past the Moon, approximated by a cylinder, for the case of the field aligned with the flow velocity, and an isotropic pressure of the magnetized plasma. He also considered qualitatively the case in which the field is slightly oblique to the flow direction for a spherical Moon (see Figure 11). The conclusions were the same as those of Michel (1968a), namely: the presence of a trailing shock wave, an umbral increase and penumbral decreases of the magnetic field magnitude.

Extending the theory of a trailing shock to very large distances behind the Moon, $> 100\ R_M$, Michel (1968b) re-interpreted the earlier IMP-1 observations and interpretations (see Subsection 2.2). He concluded that the December, 1963 data indicated the presence of too large an energy density in the magnetic field fluctuations to be caused by the lunar wake. Instead he proposed that the February, 1964 observations, with very small increases in the magnetic field fluctuations, earlier interpreted as not representing the lunar wake, could indeed be due to his proposed trailing lunar shock. The observations of Taylor *et al.* (1968) out to $5.4R_M$ do not support this view (see Section 3.1).

None of these theoretical studies by Michel, Johnson, and Midgley or by Whang or Wolf, predicted or anticipated the existence of the penumbral magnetic field increases. This feature of the magnetic field perturbations in the lunar wake was first treated by Siscoe *et al.* (1969) in a joint analysis of Explorer 35 plasma and magnetic field data. These authors found that the penumbral increases of the magnetic field are correlated with small increases of the plasma flux and possibly a small ($<3°$) outward deflection of the plasma flow away from the main ion wake (see Figure 12). (It should be noted that the digitization interval of the angular measurements of plasma flow direction for the experiment was $3°$).

From this they deduced that the effective lunar profile for the solar wind flow past the Moon included a small deflection of the plasma at the limbs which led to the de-

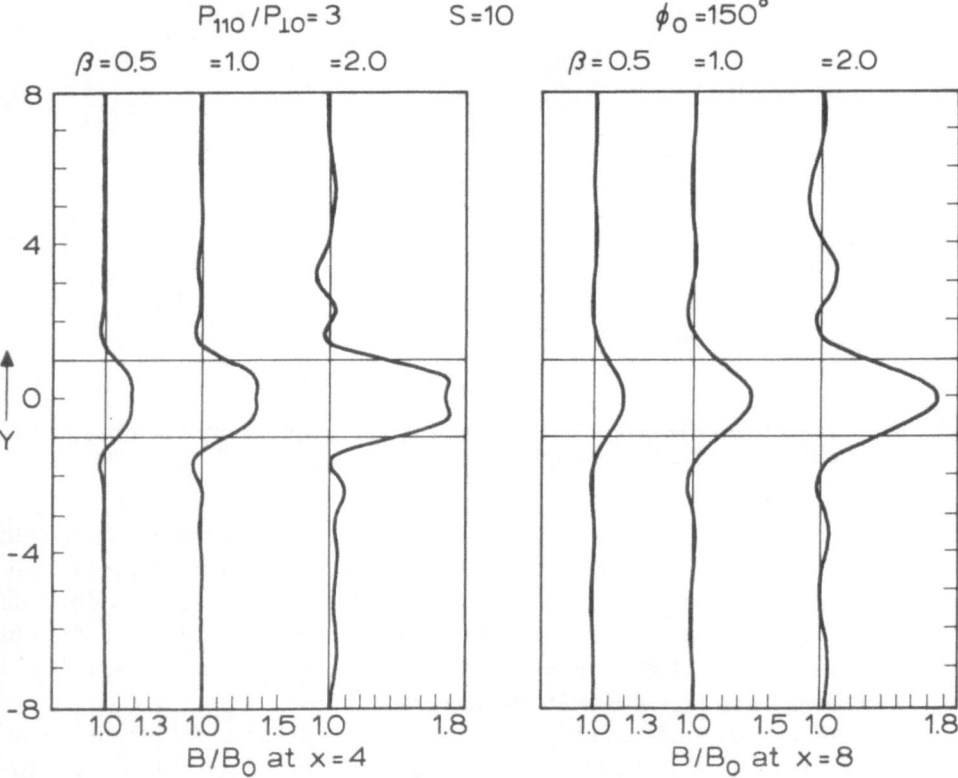

Fig. 10a.

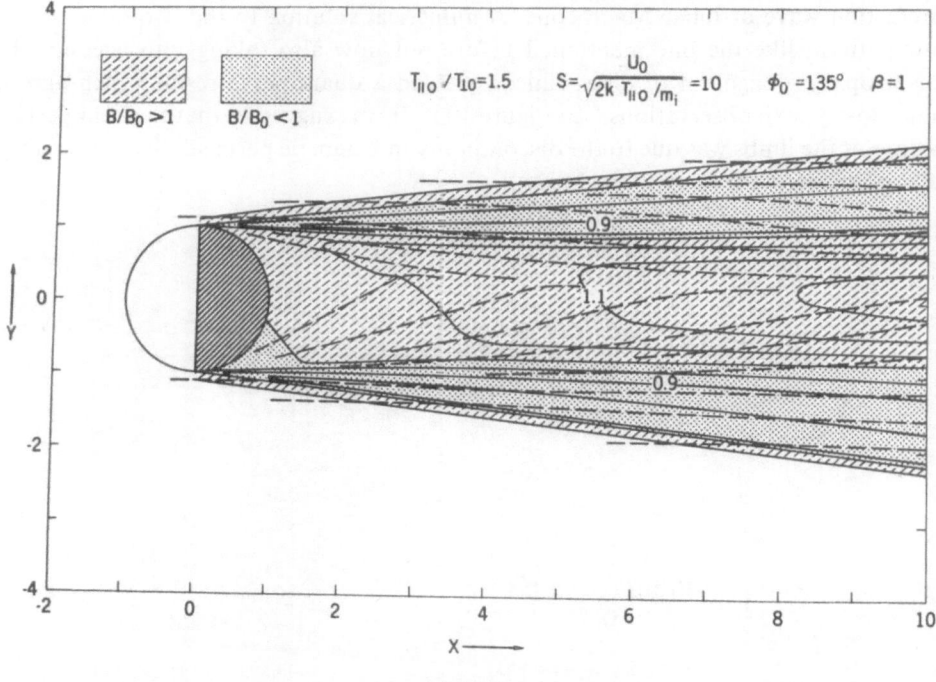

Fig. 10b.

Fig. 10a–b.   (a) Theoretical magnetic field profiles in lunar wake in the Y-direction in plane of symmetry computed from the first order approximation to the perturbed plasma ion wake (Whang, 1968b; 1969). In (b) the dashed curves represent the direction of plasma flow velocity and the shaded areas show the regions of enhancement and weakening of the magnetic field.

tached compression wave followed by the rarefaction wave and expansion fan. Three possible sources for the deflection were noted:

(1) Magnetic pressure produced electrical currents induced in the body of the Moon by the solar wind;

(2) Interaction with a neutral atmosphere; or

(3) Increased solar wind pressure due to incomplete absorption of particles that impact the surface.

In a quantitative theoretical study of the outer wake region for a two-dimensional body in the hypersonic limit, Siscoe *et al.* (1969) concluded that the plasma flow was deflected inward toward the wake axis and that the plasma flux increase was due to an increased plasma density, because the ratio of B/ϱ was constant in the penumbra of the wake. Their explanation of the penumbral increases, however, requires the introduction qualitatively of an *ad hoc* mechanism for deflection at the lunar limb.

In a modified theory that takes into account a fourth induced current source, the acceleration drift or polarization current, Whang (1969) introduced an *ad hoc* positive perturbation of the magnetic field at the limbs which propagated back outside the rarefaction wave or lunar Mach cone. A numerical solution to the two-dimensional flow pattern, like the one mentioned before but now also taking into account the anisotropic propagation of disturbances, yielded a quantitative result which agreed quite closely with observations (see Figure 10b). It was suggested that the field perturbations at the limbs was due to the discontinuity in magnetic permeability between the solar plasma and the body of the Moon.

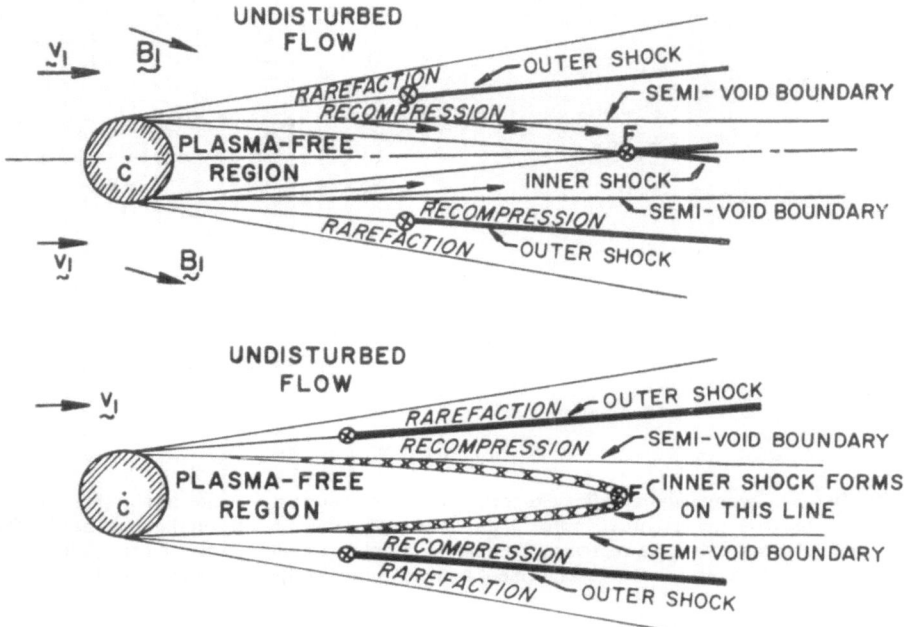

Fig. 11.   Diagram developing further geometry and flow characteristics behind the Moon leading to formation of a trailing shock wave according to Wolf (1968). The upper figure refers to the plane of symmetry while the lower figure refers to a plane perpendicular to this.

None of the theoretical studies conducted thus far is complete. Those by Whang have quantitatively taken into account the thermal anisotropy of the solar plasma although its magnitude is sufficiently small that it may not be an important parameter. He has been able to consider the case of arbitary orientation of the field relative to the solar wind velocity. The lunar perturbations can also be readily understood on the basis of fluid flow models developed qualitatively by Michel and others for the case of aligned solar wind flow and field when a three-dimensional obstacle is used to represent the Moon. Some progress towards more quantitative calculations have been made by Wolf (1968) but more work needs to be done. For both Whang and Wolf the quantitative studies have been restricted to a cylindrical approximation to the Moon so that the applicability of the results may be restricted to a region within a few lunar radii downstream of the real spherical obstacle.

No theory for the flow characteristics past the Moon was advanced before the experimental studies identifying the unique features of the magnetic field perturbations. Observations continue to provide the essential stimulus and the quantitative results for describing the solar-wind interactions. A comprehensive study of the observed characteristics of the plasma in the lunar wake has yet to be completed.

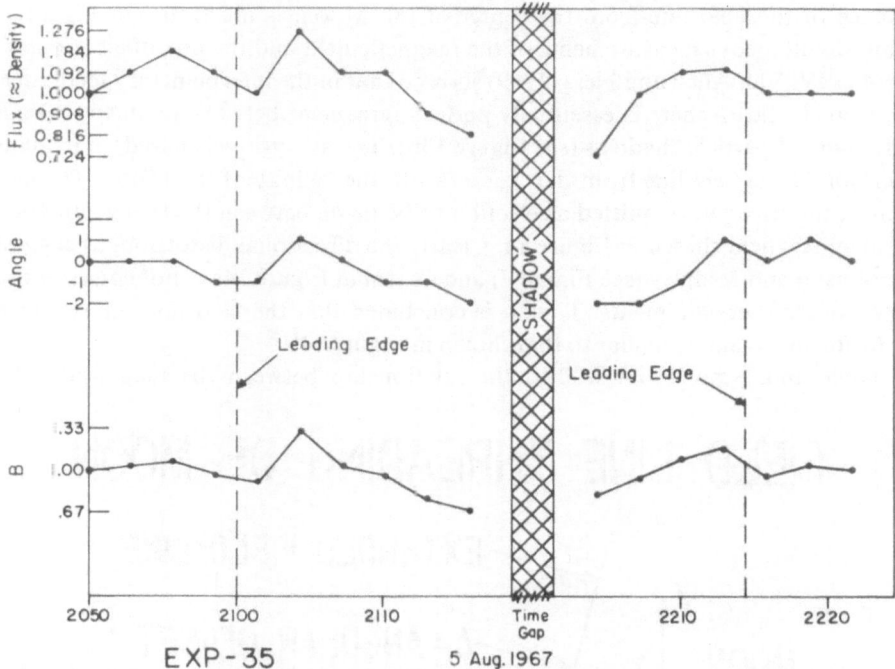

Fig. 12.   Simultaneous magnetic field data (NASA-Goddard Space Flight Center) and plasma flux data (Massachusetts Institute of Technology) showing correlated penumbral increases (Siscoe *et al.*, 1969). All data are normalized to values outside the wake.

3.3. ADDITIONAL EXPLORER 35 RESULTS

Subsequent to the early magnetic field and plasma observations discussed in Subsection 3.1, other features of the solar wind interaction were studied as well as properties of the Moon itself. In a refined study of the magnetic properties of the Moon, Behannon (1968) lowered the upper limit for the permanent dipole moment to 10^{20} G cm^3, which corresponds to a field on the surface of the Moon of less than 4γ. Also using the oppositely directed fields in the tail, he set an upper limit for the induced magnetic moment of the same value. For a homogeneous Moon, whose interior is below the Curie point, this corresponds to a magnetic permeability μ less than $1.8\mu_0$.

The absorption of particles by the Moon having energies much higher than solar wind particles was studied by Lin (1968) and Van Allen and Ness (1969). For high-energy, isotropic galactic cosmic rays, with Larmor radii much larger than the lunar

radius, these authors showed that the Moon behaves as an absorber with the occultation of particle fluxes being directly related to the solid angle subtended by the lunar
body.

For anisotropic solar electron fluxes with Larmor radii much less than the lunar
radius, the Moon behaves as an occulting disk for those field lines which intersect the
Moon. The computation of the impact parameter, D, which is a measure of the
distance of the field line from the center of the Moon, is illustrated in Figure 13a.
Using simultaneous measurements of the magnetic field and the flux of electrons with
$E_e > 45$ keV, Van Allen and Ness (1969) showed that in the interplanetary medium and
the magnetosheath there is essentially perfect agreement between predicted $(D \leqslant R_M)$
and observed particle shadows (see Figure 13b). The analysis was based on rectilinear
extension of the field line from the spacecraft to the vicinity of the Moon. The agreement of the shadows permitted a selection to be made between the two geometries of
the magnetic field shown in Figure 14. Clearly the directional distortion, as suggested
by Johnson and Midgley (see Figure 7) and shown in Figure 14a is not consistent with
the correlated measurements. Thus it is concluded that the field lines must intersect
the Moon in a manner similar to that shown in Figure 14b.

Ogilvie and Ness (1969) studied the relationship between the magnitude of the

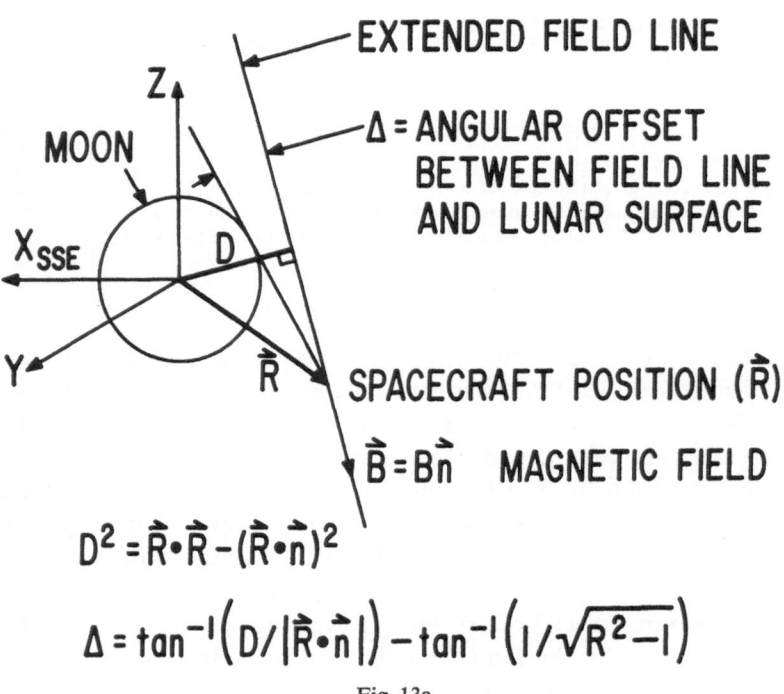

Fig. 13a.

umbral increase observed by Explorer 35 with the ion plasma characteristics observed simultaneously by Explorer 34 in interplanetary space. They determined the observed relative field increase in the umbra as a function of the β value for plasma ions. The increase was observed to be linearly dependent upon the β value as

$$\frac{\Delta |\mathbf{B}|}{B_0}\bigg|_{umbra} \doteq (0.7 \pm 0.3)\, \beta_i\,(\text{measured}). \tag{2}$$

No consistent variation with field direction was detected, although Whang's theory suggests that such a variation should be observed. In comparing the observed relationship with that predicted by Whang (1968b),

$$\frac{\Delta |\mathbf{B}|}{B_0}\bigg|_{umbra} \doteq (0.23 \pm 0.09)\, \beta_i\,(\text{theory}) \tag{3}$$

a significant discrepancy was noted. This was interpreted to be the result of not expli-

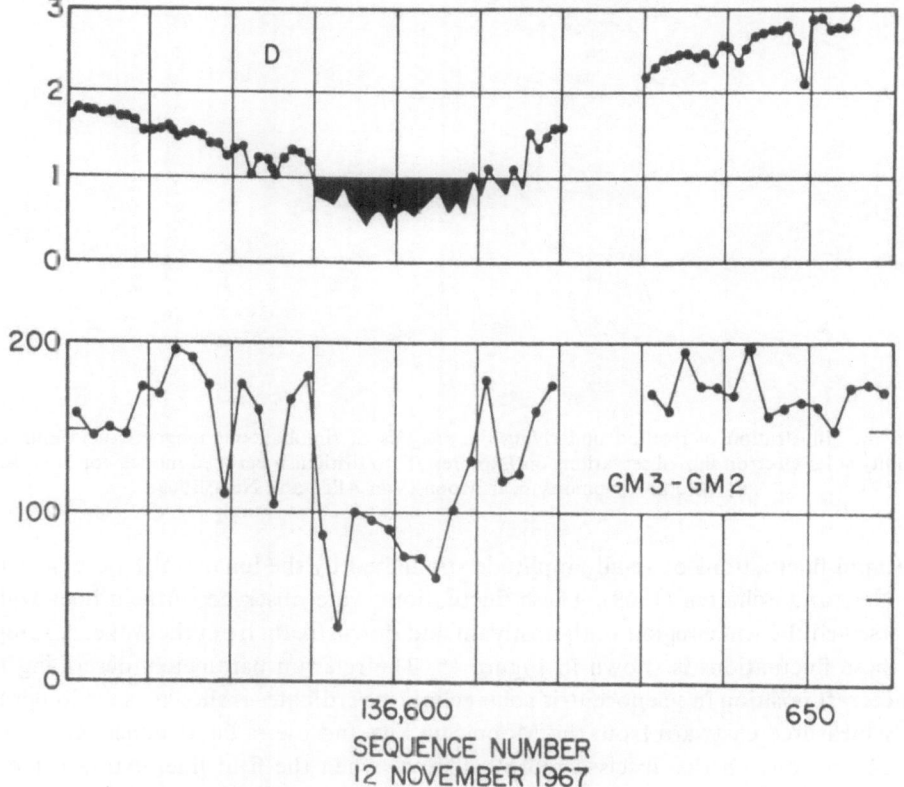

Fig. 13b.

Fig. 13a–b. (a) Geometry for the computation of the impact parameter of a field line threading the Moon and (b) experimental results from Explorer 35 using the impact parameter
(Van Allen and Ness, 1969).

citly including the electron contribution in the theory, so that these data suggested

$$\beta_{i+e} \approx 3\beta_i \qquad\qquad\qquad (4)$$

or that

$$\beta_e \approx 2\beta_i$$

in the solar wind. This use of the lunar wake as a 'solar wind sock' to measure indirectly the electron temperature in the solar wind yielded a result in favorable agreement with direct measurements.

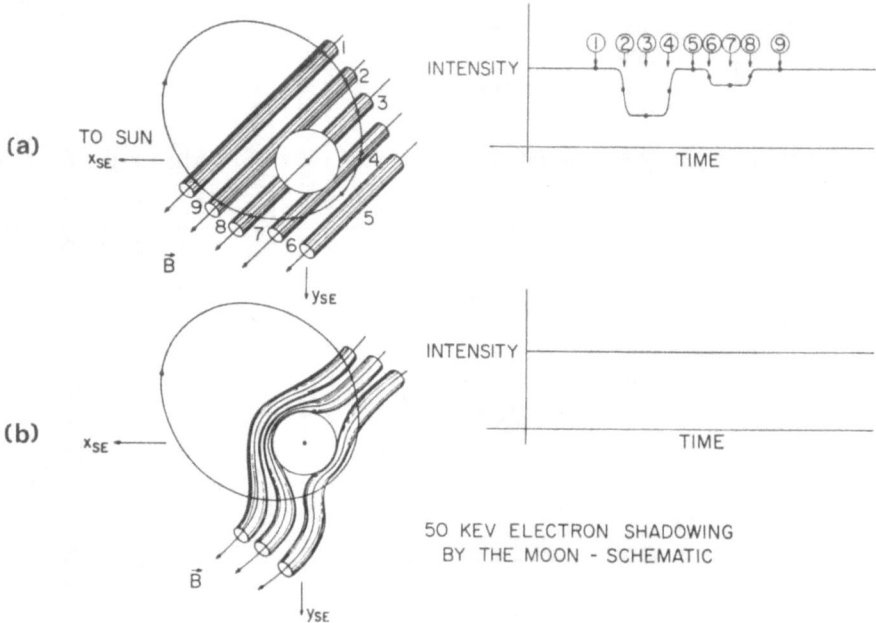

Fig. 14.   Illustration of method underlying the analysis of simultaneous magnetic field and low-rigidity solar electron flux observations on Explorer 35, to distiguish between models for interplanetary field line geometry near Moon (Van Allen and Ness, 1969).

Rapid fluctuations of small amplitude stimulated by the lunar wake were detected by Ness and Schatten (1969). These fluctuations were observed on field lines which intersected the wake region both upstream and downstream from the wake. A sample of these fluctuations is shown in Figure 15. The relevant parameters describing the spacecraft position in selenocentric solar ecliptic coordinates is shown as the longitude ϕ_{SSE} measured eastward from the Moon-Sun line and the radial distance RAD. The shaded regions on the abscissa represent times when the field line, extrapolated as shown in Figure 13a, intersected a simplified lunar wake (a cylindrical region extending downstream from the Moon). The numbers above these shaded regions indicate the distance from the spacecraft to the point of closest approach to the lunar wake along the extended field line.

It has been noted that the magnetic field is extremely steady and quiet in the core of the lunar wake.

It is seen that the noise occurs outside the umbral region in close coincidence with times when the field lines thread the lunar wake. Ness and Schatten (1969) interpret these fluctuations as associated with electrons reflected from the electric field in the penumbral region of the lunar wake. Krall and Tidman (1969) present an interpretation in terms of a ballistic wake phenomenon associated with the steep plasma density gradients in the plasma cavity boundary (the penumbra) and the occurrence of drift instabilities in such a region.

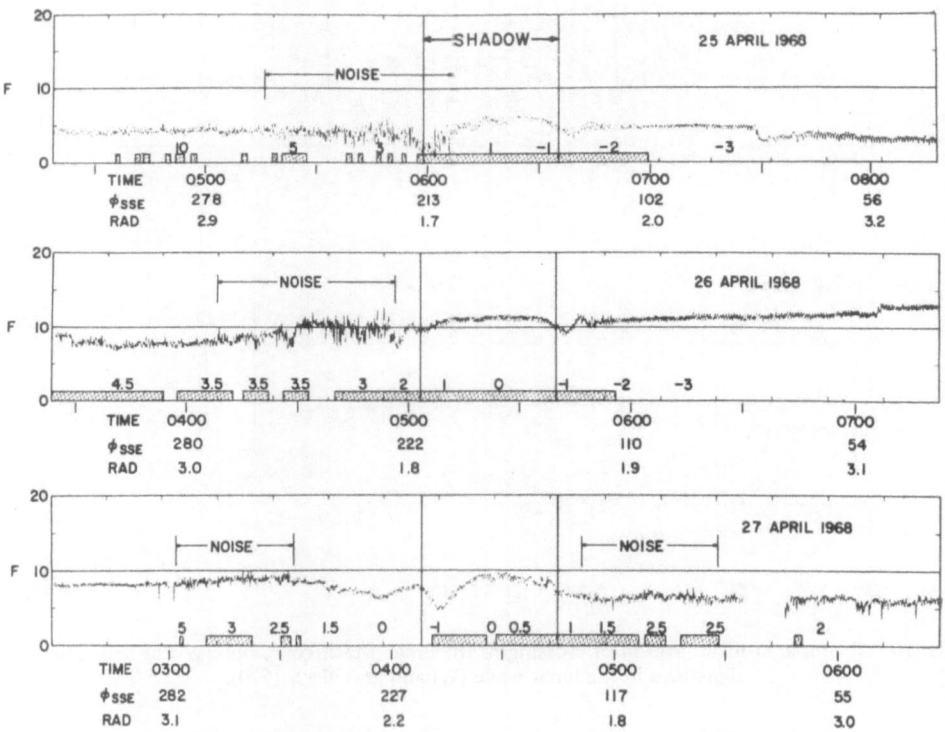

Fig. 15. Stimulated waves observed as noisy fluctuations of the magnitude of the interplanetary magnetic field, both upstream (pre-shadow) and downstream (post-shadow) from the lunar wake (Schatten and Ness, 1969). Times and satellite coordinates are given on the bottom scales.

3.4. Lunar Mach Cone

In a careful study of the geometry of the penumbral increases and decreases, Whang and Ness (1970) have confirmed the elliptical geometry of the lunar Mach cone or rarefaction wave predicated by Michel (1968a) and suggested by Johnson and Midgley (1968). The method of identifying the Mach cone using the NASA-GSFC magnetic field data is illustrated in Figure 16. It is assumed that the change in sign of the magnetic field anomaly in the penumbra coincides with crossing of the rarefaction wave or Mach cone. When the exterior penumbral positive anomaly is not present the crossing is defined by the outer limit of the penumbral decrease. Using the spacecraft position relative to the plane of symmetry, as shown in Figure 17, it is then possible to relate the Mach angle, α, to the angle ψ between the direction of propagation of the rarefaction wave **K** and the magnetic field **B**$_0$.

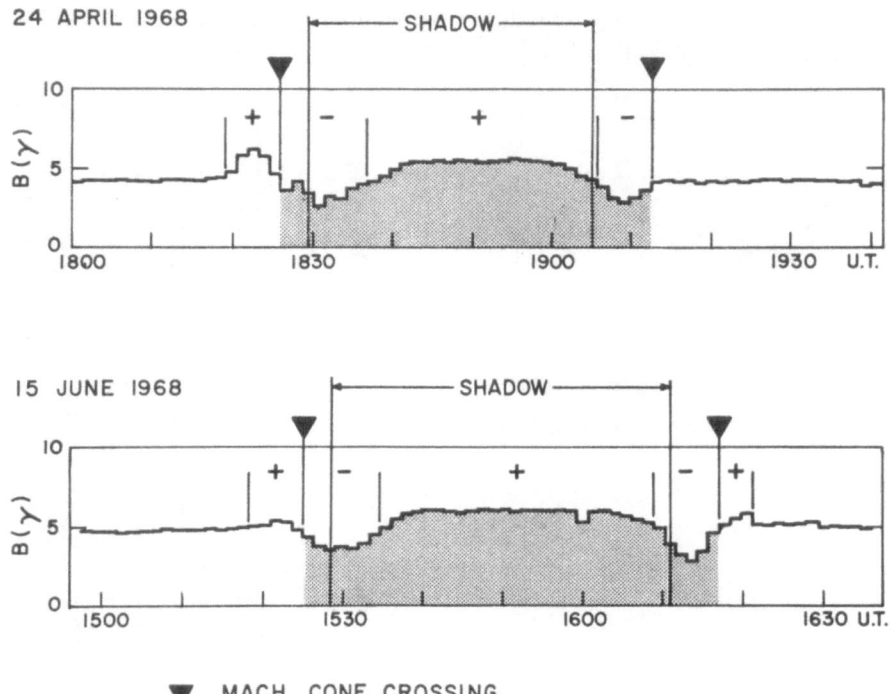

Fig. 16.   Example of identification of crossing of the lunar Mach cone, observed in magnetic field signatures in the lunar wake (Whang and Ness, 1970).

Recalling that the orbital plane is almost parallel to the ecliptic plane permits a measurement of the aberrated solar wind flow direction. Determination of the Mach angle α as a function of the angle ψ provides the first measurements of the anisotropic propagation of magnetoacoustic waves in the solar plasma. The experimental results for $\psi = 90°$ are shown in Figure 18, where it is seen that the axis of the Mach cone deviates from the Moon-Sun line by 4.5°. The solar plasma speed measurements made by the MIT group with Explorer 35 (H. S. Bridge and J. H. Binsack, private communication) during the same time interval predicts an aberration of $4° \pm 0.5°$. The aberration could be due to an azimuthal component of the solar wind velocity of 5 ± 5 km/sec at 1 AU opposite to the direction of planetary motion.

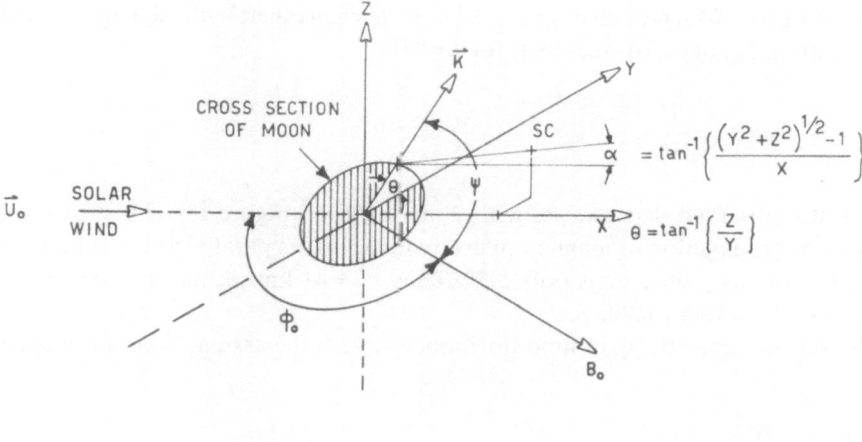

$$\cos \psi \approx \sin \phi_0 \cos \theta$$

Fig. 17. Geometry for the analysis of the variation of lunar Mach cone angle, α, relative to the spacecraft position (SC), the field line orientation, (B_0) and plane of symmetry. The propagation angle, ψ, is defined as shown.

ENTERING SIDE LEAVING SIDE

○ FIELD INCREASE △ FIELD INCREASE
● FIELD DECREASE ▲ FIELD DECREASE

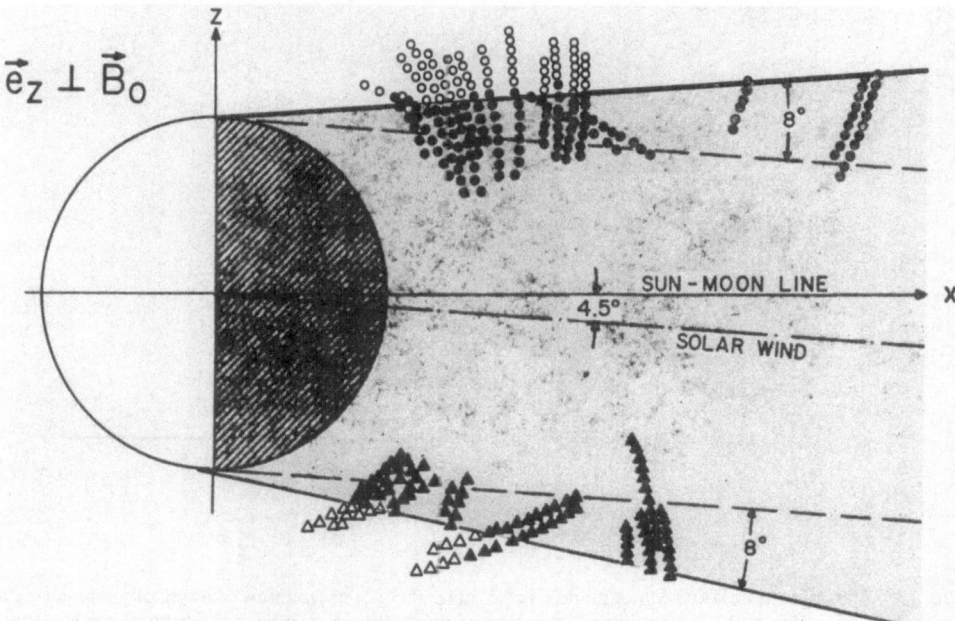

Fig. 18. Summary of Mach cone crossings corresponding to propagation angle, ψ, of approximately 90° (Whang and Ness, 1970).

The observed Mach cone angle, α, of 8° yields a magnetohydrodynamic velocity of propagation, $V_\perp$, of 60 km/sec since, for $\psi = 90°$,

$$\sin \alpha_\perp \approx \frac{V_\perp}{V_{sw}}. \tag{5}$$

It is also observed that α varies with ψ, as shown in Figure 19. This shows that the velocity of propagation of magnetohydrodynamic waves parallel to the field direction, $V_{\parallel}$, is less than $V_\perp$ since $\alpha_{\parallel}$ is only 5.5°. Thus $V_{\parallel} = 41$ km/sec and the velocity aniso-tropy is $V_\perp / V_{\parallel} = 1.45 \pm 0.20$.

The velocity anisotropy in fluid flow models, with the assumption of a magnetized

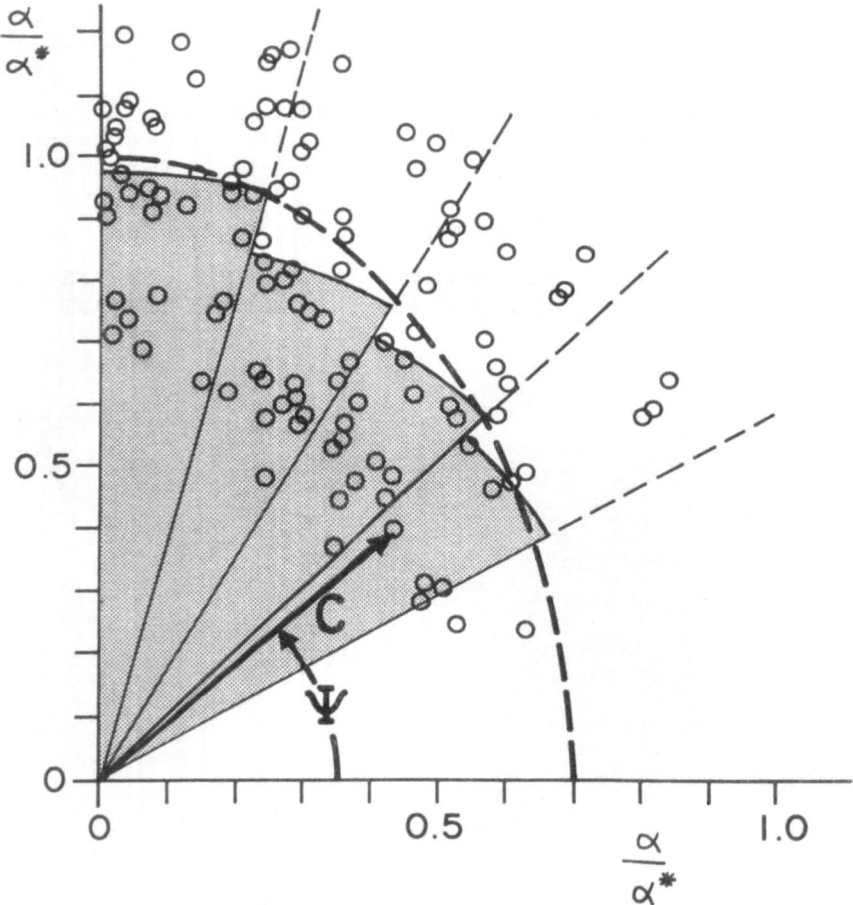

Fig. 19.   Variation of relative Mach angle (α/α^*, where $\alpha^* = 8°$) as function of angle of propagation, ψ, as defined in Figure 17. The elliptical variation of (α/α^*) with ψ is evidence for anisotropy of the propagation velocity of magnetoacoustic waves in the solar wind (Whang and Ness, 1970).

fluid is given by:

$$\frac{V_{MA}}{V_S} = \sqrt{\frac{\gamma p + B^2/4\pi}{\gamma p}} = \sqrt{1 + \frac{2}{\gamma \beta}}. \tag{6}$$

For a magnetized collisionless plasma, Lüst (1959) has shown that

$$V_\perp = \sqrt{\frac{1}{\varrho}(2p_\perp + B^2/4\pi)} \tag{7}$$

$$V_\| = \sqrt{\frac{1}{\varrho}(p_\perp - p_\| + B^2/4\pi)}. \tag{8}$$

Thus, the velocity anisotropy for this more realistic model of the interplanetary medium is:

$$\frac{V_\perp}{V_\|} = \sqrt{\frac{1 + \beta}{1 - \beta(\eta - 1)/2}}. \tag{9}$$

Using nominal values of $\beta = 1$, $\eta = 1.2$ and $\gamma = \frac{5}{3}$ we obtain:

$$\frac{V_{MA}}{V_S} = 1.48 \quad \text{and} \quad \frac{V_\perp}{V_\|} = 1.49. \tag{10}$$

These are in reasonably good agreement with the observational result of 1.45. The discrepancy between theory and observation can be accounted for in the fluid flow model by either increasing γ or β. For the magnetized collisionless plasma theory, a reduction of β or η is sufficient. Since the electrons are probably much more isotropic than the ions and are much hotter, they act to reduce the value of η.

It should be noted, of course, that the absolute values of the velocities depend on the use of the supporting plasma data and assume that there are no explicit correlations between solar plasma speed and direction. This later assumption is probably not completely valid but it does not alter the interpretation of the observed velocity anisotropy of 1.45.

4. Lunar Electrical Conductivity

As discussed by Gold (1966) and Tozer and Wilson (1967), the electrical conductivity of the Moon's interior is the principal factor determining the characteristics of the solar wind interaction with the unmagnetized atmosphereless Moon. Thus, by inverting this relationship, it should be possible to deduce characteristics of the interior electrical conductivity from the observed features of the lunar wake. Since the observations are indirectly related to the electrical conductivity, it is necessary to have a number of models of possible conductivity structure available for comparison.

The electrical conductivity of typical minerals and rocks in the Earth's mantle depends on temperature, pressure, composition, and the fluid filling the cracks, pores and interstices. At the elevated temperatures and pressures in the deep mantle (>400 km) only the pressure, temperature, and composition are important. In the case of the Moon, devoid of an atmosphere and thus probably also of fluids in its surface layers, the principal parameter affecting the electrical conductivity will be the temperature. The maximum pressure is not high enough to affect the conductivity to an appreciable degree.

The electrical conductivity of typical silicate minerals can be well approximated by:

$$\sigma = \sum_{i=1}^{2} \sigma_i \exp(-E_i/kT), \tag{11}$$

where T is the absolute temperature in degrees Kelvin, k is the Boltzmann constant, and the coefficients σ_i and activation energies E_i depend on pressure and composition. The two mechanisms for electrical conduction are intrinsic semi-conduction and ionic conduction. Using Equation (11) and appropriate values for σ_i and E_i, England et al. (1968) have derived a number of conductivity models based on other independent temperature distributions derived from computations of the thermal history of the Moon. For this work, they used values of σ_i and E_i for olivine, an iron-magnesium silicate with

$$\left.\begin{array}{l} \sigma_1 = 5.5 \times 10^1 \text{ mho/m} \\ E_1 = 0.92 \text{ eV} \end{array}\right\} \text{ for intrinsic semi-conduction} \tag{12}$$

and

$$\left.\begin{array}{l} \sigma_2 = 4 \times 10^7 \text{ mho/m} \\ E_2 = 2.7 \text{ eV} \end{array}\right\} \text{ for ionic conduction}.$$

Also,

$$\frac{\partial E_1}{\partial p} = -5.1 \times 10^{-3} \text{ eV/kb}$$

$$\frac{\partial E_2}{\partial p} = +4.8 \times 10^{-3} \text{ eV/kb}.$$

The pressure at the center of the Moon is approximately 50 kb, so that the principal mechanism for electrical current flow is ionic conduction in these models.

A summary of their results for the radial variation of σ, with the assumption of a Moon of homogeneous composition is shown in Figure 20a for different thermal models ranging from young (0.9 billion years, curve A), to old (4.5 billion years curve B), to older yet, with melting (Curve C). It is seen that the conductivity is very low in the outermost layers of the Moon, increasing rapidly until a depth of ≈ 400 km is reached. At greater depths the conductivity increases much less rapidly; but there are significant differences in the values in the deep interior depending upon the assumed age of the Moon, because the thermal history models used by England et al. (1968) used radioactive heat sources with the same composition as chondritic meteorites.

The corresponding temperature profiles for the conductivity curves are shown in Figure 20b. Here the cause of the sharp increase of conductivity with depth in the near surface layers and the eventual limiting values for the core is evident. Using slightly different values of σ_i and E_i and a different lunar thermal history model, Tozer and Wilson (1967) derived a conductivity profile similar to curve C in Figure 20a. The very low thermal diffusivity of silicates essentially traps all the heat generated below approximately 400 km in the Moon.

The marked dependence of lunar electrical conductivity on temperature is a relationship which can also be inverted so that given the electrical conductivity of the Moon and its composition, an estimate of its internal temperature can be made. It should be noted that for the general range of temperatures in the lunar interior, a change by a factor of 10 in conductivity corresponds to a 200 K change in temperature for temperatures between 1000 K–2000 K. The relationship (11) between temperature and conductivity depends on composition such that a sequence of typical materials described geologically as being the same in fact yields a variation by one order of magnitude larger or smaller in σ. Thus it is clear that the variation of composition will lead to internal temperature estimates with an uncertainty of ± 200 K, even if the electrical conductivity is known precisely.

In their interpretation of the early Explorer 35 measurements, Ness *et al.* (1967) used arguments similar to those of Gold (1966) and Tozer and Wilson (1967) regarding the diffusion time of magnetic field lines in the lunar interior in order to deduce the effective electrical conductivity. For the transient case, the characteristic diffusion time, τ_D, the 'Cowling' time constant, is

$$\tau_D = \mu\sigma L^2, \tag{13}$$

where L is the characteristic scale length of the body in the one-dimensional model. The absence of a bow shock was interpreted as indicating that the value of τ_D must be less than the time for transit of the field lines carried by the solar wind past the Moon.

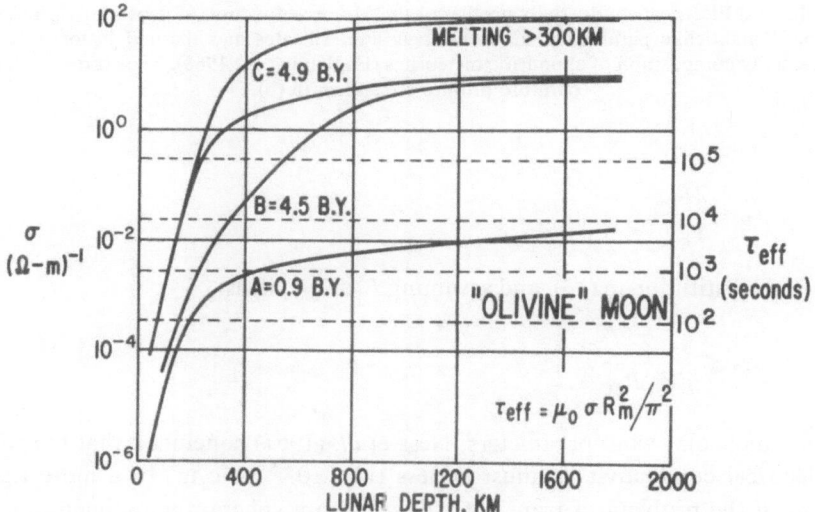

Fig. 20a.

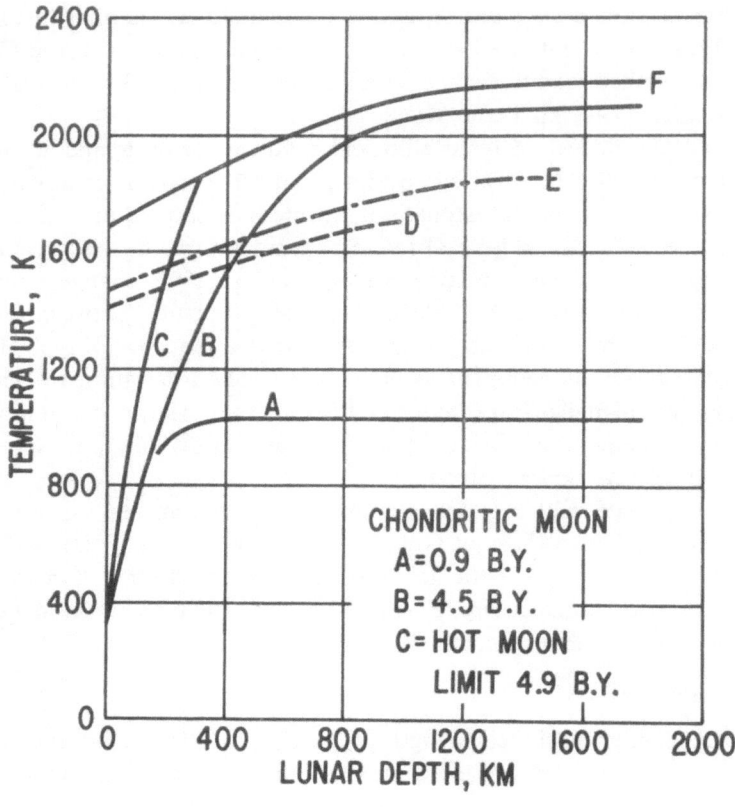

Fig. 20b.

Fig. 20a–b. (a) Electrical conductivity profiles of the Moon as function of depth for various models using typical mantlelike parameters for iron-magnesium silicates and thermal histories based on the radioactivity composition of chondritic meteorites (England *et al.*, 1968). The corresponding temperature profiles are shown in (b).

Thus

$$\tau_D \leqslant \frac{2R_M}{V_{sw}},$$ (14)

which after substitution in (13), and assuming $L \approx R_M$, yields

$$\sigma_{\text{eff}} \leqslant \frac{2}{\mu V_{sw} R_M}.$$ (15)

Using nominal solar wind parameters, Ness *et al.* (1967) concluded that the effective lunar electrical conductivity σ must be less than 10^{-5} mho/m. In a more rigorous treatment of the problem of transient induction in a sphere, three-dimensional geometry introduces a modification of the time constants so that the Cowling diffusion time constant for the mode of lowest order is given by:

$$\tau_D = \mu \sigma R^2 / \pi^2 , \tag{16}$$

With the assumption that the scale length $L = R_M$ for a homogeneous Moon, the characteristic time constants corresponding to various values of the conductivity are also given in Figure 20a.

While the computed conductivity profiles show that the assumption of a homogeneous Moon is not valid, they do indicate that the values for σ are many orders of magnitude higher than can be considered consistent with an observed time constant of less than 10 sec. From this, Ness *et al.*, (1967) taking into account the three-dimensional geometry, concluded that the average temperature of the Moon must be less than 1000 K. Colburn *et al.* (1967) supported the conclusion of a low bulk electrical conductivity for a homogeneous Moon. However, they believed that attempts to determine an interior temperature on the basis of this effective conductivity were premature: the inhomogeneous Moon might produce no bow shock but still have a highly conductive interior. The next subsection discusses this possibility in detail.

4.1. UNIPOLAR INDUCTOR: STEADY STATE

As the solar wind convectively transports the embedded interplanetary magnetic field past the Moon, it generates an electric field in a coordinate system fixed to the Moon. At 1 AU, the magnitude of this field is given by

$$\mathbf{E} = - \mathbf{V}_{sw} \times \mathbf{B}_0 \tag{17}$$

and amounts to $\approx 1 \mathrm{mV/m}$ for typical solar wind parameters. This electric field, unless opposed by a polarization of the Moon as a dielectric, will lead to electrical currents which generate a secondary magnetic field. This induced current system and the effect it would have on the solar wind flow was first considered quantitatively by Sonett and Colburn (1967) for a homogeneous Moon, and by Hollweg (1968) and Sonett and Colburn (1968) for an inhomogeneous Moon.

The reality of the unipolar mechanism depends upon the ability of the electrical current to form a closed circuit by means of a return path in the solar plasma. Although the exact nature of the plasma-Moon interface is not yet known, it has been assumed that the equivalent electrical impedance is much less than that associated with the current flow in the lunar interior. It is of course well known that the conductivity of the solar plasma is very high. Indeed, it is this characteristic of the solar plasma which leads to the 'frozen-in' magnetic field and the unipolar induction mechanism itself.

The quantitative study by Hollweg (1968) assumed the Moon to be a simple sphere with two concentric layers representing the core and crust. Using as the inducing field Equation (17), he derived extreme values of the conductivity of the core and crust, for limiting cases of the ratio of conductivities in the crust and core. The extreme value was based upon the absence of plasma flow deflection, according to Equation (1), as the necessary condition determining the presence or absence of a bow or limb shock wave. It should be noted that in both the work of Hollweg (1968) and that of Sonett and Colburn (1968) the secondary fields were computed with a cylindrical Moon although the current flow was based on a spherical Moon.

Hollweg's results for the steady-state flow may be summarized as follows:

(1) For a conducting thin crust, supporting the major fraction of the current, the interior conductivity must be much less than 10^{-5} mho/m.

(2) For a crust of low but finite conductivity, with the core supporting the major fraction of the current, the core conductivity must be much less than 2.6×10^{-5} mho/m.

(3) For a thin insulating crust, shielding a conductive core, the crust conductivity must be less than 10^{-10} mho/m for a crust 10 m thick, and 10^{-7} mho/m for a crust 10 km thick.

Since conductivities as low as those required in Case (3) seemed unlikely, Hollweg (1968) concluded that the presence of a conducting core (i.e. $\sigma_c > 10^{-5}$ mho/m) was highly unlikely.

For non-steady flow, Hollweg (1968, 1970) introduced the concept of 'conducting islands', that is, localized disk-like regions near the surface, having relatively high conductivity say, 10^{-3}–10^{-4} mho/m. The strength of secondary magnetic fields induced by fluctuations of the interplanetary magnetic field would be high enough to cause deflection of the solar wind flow at the limbs of the Moon and thus explain the observed penumbral increases in the interplanetary magnetic field discussed in Sub-section 3.1. These 'conducting islands' thus represent a physical basis for the suggestion by Siscoe *et al.* (1969) of an *ad hoc* mechanism for solar wind flow deflection at the lunar limbs. Hollweg noted that limb shocks thus generated by the unipolar induction mechanism would form only in the vicinity of the plane defined by the field and solar wind velocity vector. The conducting-islands' concept does not suffer from such a limitation.

Schwartz *et al.* (1969) have repeated the earlier study by Sonett and Colburn (1968) on the unipolar induction mechanism, extending it to include several different thermal profiles and conductivity functions. They conclude that a Moon with a hot interior but cool exterior can cause the reported plasma deflection at the lunar limb by the unipolar mechanism. This will generate the observed limb shock wave, which is their interpretation of the phenomenon of penumbral increases. While they do not give the conductivity profile for their preferred model, which is based on a diabase composition, the relationship of conductivity to temperature for the composition and conductivity of this model is given:

$$\sigma_D = 10^3 \exp\left(-0.634/kT\right). \tag{18}$$

The internal temperatures in all their models is approximately 1700–1900 K for depths greater than 400 km, so it is clear that Schwartz *et al.* (1969) predict a hot lunar interior with an electrical conductivity of ≈ 10 mho/m.

The results of Whang and Ness (1970) on the lunar Mach cone geometry are especially significant in the framework of interpretations of the magnetic field penumbral increases and the plausibility of the lunar unipolar induction mechanism. The induced magnetic field arising from the unipolar mechanism is toroidal, with the axis of symmetry perpendicular to the plane of symmetry defined by $\mathbf{V}_{sw}$ and $\mathbf{B}_0$. The magnitude of the field, using the symbols defined in Figure 17, is related to the magnetic field, solar wind velocity, and angle θ as follows:

$$H_\phi \propto \mathbf{B}_0 \times \mathbf{V}_{sw} \cos \theta, \tag{19}$$

so that the magnetic back pressure for plasma deflection is

$$P_{\text{MAG}} \propto V_{sw}^2 B_0^2 \sin^2 \phi_0 \cos^2 \theta. \tag{20}$$

The solar wind pressure normal to the lunar surface (and the toroidal magnetic field) depends upon the angle Δ of the normal measured with respect to the limb of the Moon as:

$$P_{sw} \propto n V_{sw}^2 \sin^2 \Delta. \tag{21}$$

Thus for deflection to take place we equate (20) and (21) to obtain:

$$\sin \Delta \propto \frac{B_0}{\sqrt{n}} \sin \phi_0 \cos \theta \tag{22}$$

relating the deviation of the flow and the properties of the magnetized solar wind and the position of the observation point relative to the plane of symmetry.

However, $\sin \theta_0 \cos \theta = \cos \psi$, the angle of propagation used by Whang and Ness (1970) in their Mach cone geometry study (see Subsection 3.4). Thus, for small angles of deflection where $\sin \Delta = \Delta$, one finally obtains:

$$\Delta \propto \frac{B_0}{\sqrt{n}} \cos \psi. \tag{23}$$

The Mach cone, as defined and analyzed by Whang and Ness (1970), is the surface separating penumbral increases and decreases in the lunar wake. It is reasonable to assume that this surface would be the same or linearly related to the surface defined by the deflected plasma. In this case Δ is a maximum at $\psi = 0^0$ and a minimum at $\psi = 90^\circ$.

It will be recalled from Figure 19 that the observed variation of the Mach cone angle α with ψ is elliptical, with a maximum at $\psi = 90^\circ$ and a minimum at $\psi = 0^\circ$. This is exactly the opposite relationship from that to be expected if unipolar induction is an important mechanism affecting the solar wind interaction with the Moon and responsible for the penumbral increases. Examples of clearly observable penumbral anomalies for the case of aligned field and flow are shown in Figure 21. Under the assumption that the unipolar induction mechanism is responsible for penumbral increases, they should not be observed for $\theta_0 = 0^\circ$ (or 180°). Their existence provides further evidence that some mechanism other than plasma deflection by induced magnetic fields must be responsible for the observed penumbral increases.

A possible source, which has yet to be considered, is the effect of the finite Larmor radius of the solar wind ions and the near coincidence of their cyclotron period with the convection time past the Moon. Clearly, for such a small-scale effect as the penum-

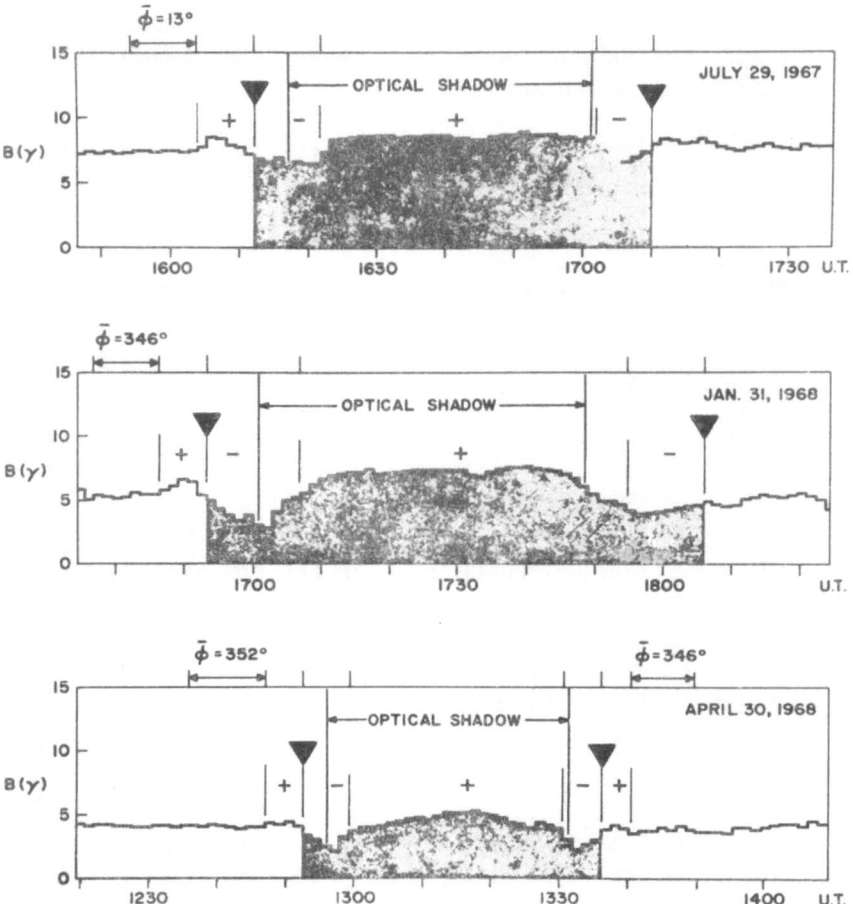

Fig. 21. Observation of identifiable penumbral increases in the interplanetary magnetic field as observed by the NASA-GSFC magnetic field experiment when the field direction is almost aligned with solar wind velocity.

bral increases imply in the overall solar wind interaction problem, this feature of the flow must be considered in future studies.

4.2. TIME-VARYING INDUCTION

Another method to investigate the electrical conductivity of the lunar interior, different from that of studying the steady state characteristics of the solar wind interaction discussed so far in Sections 4 above, is to employ the naturally occurring fluctuations of the interplanetary magnetic field. In studies of the conductivity of the deep terrestrial mantle, large quasi-sinusoidal variations with periods of days and weeks driven by ionospheric current systems are involved. These primary currents induce secondary currents in the Earth by Faraday's Law; the magnetic field associated with the currents is used to deduce the nature of the subsurface electrical conductivity. Naturally occurring electromagnetic signals at much higher frequencies generated by lightning have been used to study the shallow mantle conductivity.

In the case of the Moon, the interplanetary medium moves supersonically past the body so that all explicit time variations observed in a coordinate system fixed to the Moon are directly related to the spatial variation of the interplanetary medium rather than its temporal variations. There exists little evidence for the existence of waves with quasi-sinusoidal variations in the solar wind. Rather the interplanetary medium appears to be micro-structured into a large number of regions of magnetized plasma, separated from each other by classical magnetohydrodynamic discontinuities, which are convectively transported outward from the Sun. The discontinuities are most clearly observed as very sudden changes in the interplanetary magnetic field (and plasma) which occur in less than 10–100 sec. In order to use such discontinuities, the three following conditions must be satisfied:

(1) Explorer 35 must be located within the plasma umbra or penumbra so that secondary fields can be measured;

(2) Independent simultaneous observations in cislunar space of the time profile of the discontinuity must be made, in order to establish a reference scale for the changes; and

(3) A discontinuity with a short time constant, < 10 sec, must be convected past the Moon at such a time that it can be observed by Explorer 35.

Ness (1968, 1969) has used these discontinuities to obtain an independent estimate of the conductivity of the lunar interior.

Simultaneous data from Explorers 33 and 35 which satisfy the above criteria are shown in Figure 22. The two events were selected from a total of six which were the only ones observed during the period July 1967–June 1968. In both cases the field change associated with the discontinuity surface is observed principally as a change in the selenocentric latitude, θ_{SE}, of the field vector. For each event, both satellites, using 81.8-sec averages of the measured magnetic field, observed nearly identical discontinuities. Superposed are the steady-state perturbations of the magnetic field in the lunar wake consisting of the indicated umbral and penumbral anomalies. It should also be noted that during the several hours of interplanetary magnetic field data, the two satellites measure essentially the same magnetic field, except for the time offsets noted on the scales at the top. This is a characteristic feature of the interplanetary medium, that it is uniform on the scale of cislunar distances except for the embedded planar discontinuity surfaces.

A more detailed time profile of these two events based on fine time-resolution data obtained at 5.11-sec intervals is presented in Figure 23. This shows that for the February event there is a measurable time dilation of the discontinuity from less than 10.2–56.2 sec observable in both the X and Z components, but for the May event the time profile is identical within the resolution of the data: 40.9 and 36.8 sec.

The first event is interpreted in terms of the Cowling diffusion time constant for the lowest-order mode being less than $\frac{1}{4}$ of the 56.2 sec time interval during which the

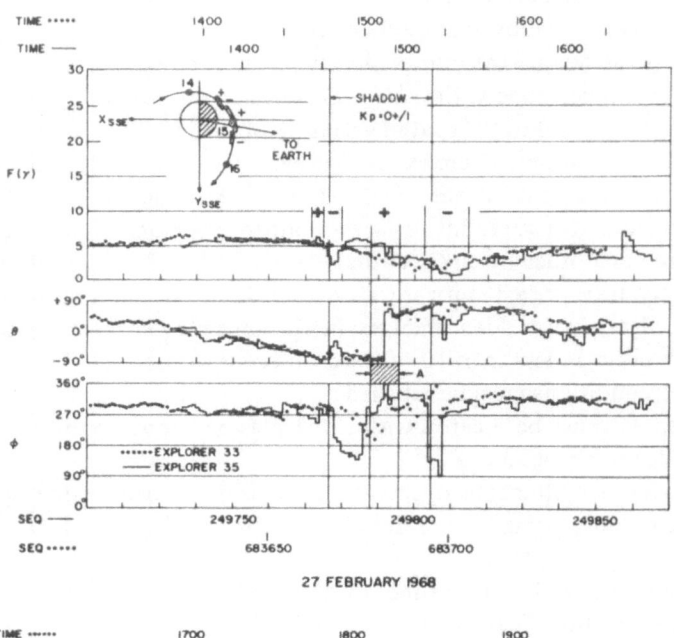

27 FEBRUARY 1968

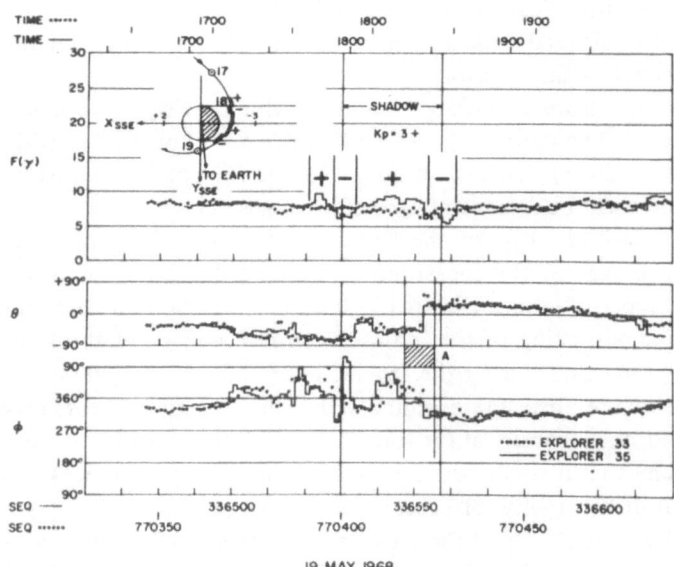

19 MAY 1968

Fig. 22.   Simultaneous Explorers 33 and 35 measurements of the interplanetary magnetic field as a discontinuity passes the Moon at times when the latter spacecraft was located in the lunar wake region (Ness, 1969). Dots refer to Explorer 33 and solid lines to Explorer 35.

eddy currents induced in the Moon and their magnetic fields decay to zero. This value of $\tau_D < 20$ sec compares very favorably with that determined independently from the steady-state flow (see beginning of Section 4). The assumption of a two-layer Moon with negligible crustal conductivity and a core radius of approximately 1350 km yields an interior conductivity of 8×10^{-5} mho/m. This corresponds to a temperature estimate of 800 ± 200 K using typical values for olivine. These results were first reported by Ness (1969).

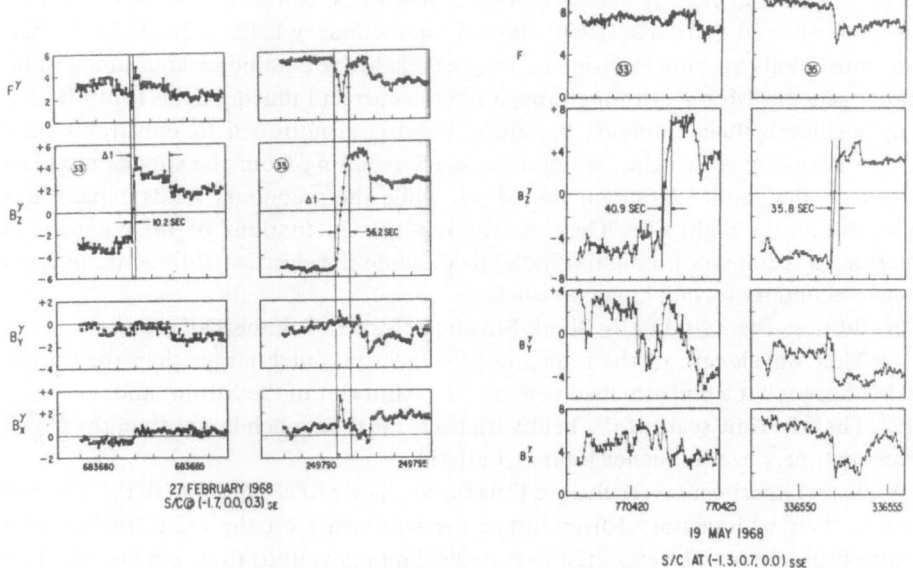

Fig. 23. Detailed magnetic field data on an expanded time scale (by factor of 16 from that used in Figure 22), showing the effect of the lunar body in distorting the time profile when the spacecraft is in core of the wake (a), and the lack of effect when the spacecraft is in the penumbra (b) (Ness, 1969).

The second event in May 1968 shows no similar time dilation effect. This is believed to be due primarily to the long time-width of the discontinuity (> 10 sec) and to the distance of the satellite from the umbra. At the time of observation it was located $0.7\ R_M$ from the axis of the cavity, while for the February 1968 event; it was only $0.3\ R_M$ from the axis. The induced fields of the Moon are small at satellite altitude and confined to the interior of the plasma cavity. When Explorer 35 is located either in the penumbra or far downstream from the Moon, there is no measurable effect on discontinuity time profiles which can be associated with the lunar interior (Ness, 1968). There is as yet no theory for the transient response of the Moon to discontinuities in the interplanetary medium.

Induced fields in the Moon due to sinusoidal variations of the external magnetic (and/or electric) field have been considered by several authors with certain simplifying assumptions. Blank and Sill (1969a) studied the secondary magnetic fields induced in a homogeneous Moon by a sinusoidally varying magnetic field with the assumption that the secondary magnetic fields were axially symmetric and confined to meridian planes. They failed to include the motional electric field induced by the flow of the solar wind past the Moon and also used the wrong boundary conditions. The appropriateness of their analysis was questioned by Fuller and Ward (1969); and Blank and Sill (1969b) replied to their critique. Subsequently Sill and Blank (1970) presented a revised analysis of the problem, taking these criticisms into account and also adding the refinement of a radially inhomogeneous Moon.

In a similar and slightly earlier sequence of studies, Schwartz and Schubert (1969) and Schubert and Schwartz (1969) studied the secondary fields induced in the Moon by a sinusoidally varying electric and magnetic field for a homogeneous and a radially inhomogeneous Moon. Although they used the correct inducing fields (both **B** and **E**) in their studies, their boundary condition imposed a non-realistic constraint since it did not correctly reflect the strong asymmetry resulting from the supersonic plasma impact on the sunlit surface of the Moon while the secondary fields expand into a a vacuum on the night side. They showed that the fluctuations of the magnetic field generated a secondary poloidal (dipole) field, while the electric field fluctuations generated a secondary toroidal magnetic field.

In addition, these studies by Blank-Sill and Schwartz-Schubert all require:

(1) That wavelength of the inducing field must be much larger than the radius of the Moon, so that a uniform field penetrates the interior of the Moon; and

(2) That the time scale of the field variations must be much larger than the Cowling diffusion time, i.e. a frequency limit $f \ll 0.01$ Hz.

While no experimental results are thus far available to compare with these theories, even in their rudimentary form, future measurements on the lunar surface in the Apollo program may be expected to provide data relevant to these studies when compared to simultaneous measurements in orbit from Explorer 35. Until the boundary value problem is correctly formulated, however, such data cannot be uniquely interpreted.

It should be noted that these theories assume *a priori* that the magnetic fluctuations (and their associated electric fields) uniformly excite the lunar interior. It is by no means clear, however, that this requirement is met for the Moon because, as the plasma is absorbed on the sunlit hemisphere of the surface, the fluctuations are converted to electromagnetic waves. But on the dark side of the Moon, there is no corresponding excitation, so that there is hemispherical asymmetry of the source. On the sunlit side, the normal component of the secondary magnetic field must be zero since the conductivity of the solar plasma is very high. On the night side, however, only the divergence-free constraint on **B**, namely $\mathbf{B}_I \cdot \mathbf{n} = \mathbf{B}_E \cdot \mathbf{n}$, is required where $\mathbf{B}_E$ is the external field and $\mathbf{B}_I$ the internal field, and **n** is the normal to the lunar surface. Thus there is also a strong asymmetry in the boundary conditions.

The net inducing field is thus rather substantially modified from that assumed by Blank, Sill, Schwartz and Schubert. Although the problem of the electromagnetic response of a conducting sphere in a time-varying electromagnetic field is an old classical problem (for example, see Debye (1908) and Wait (1951)), the case of the Moon is sufficiently different that the simple modifications thus far studied are probably not appropriate to represent the real problem.

5. Summary

Early satellite studies related to the solar wind interaction with the Moon before 1967 did not correctly reveal the characteristic features of the phenomenon. The U.S.A. Lunar Explorer 35 has provided since July, 1967 definitive experimental results regarding the perturbations of the interplanetary magnetic field and plasma in the lunar

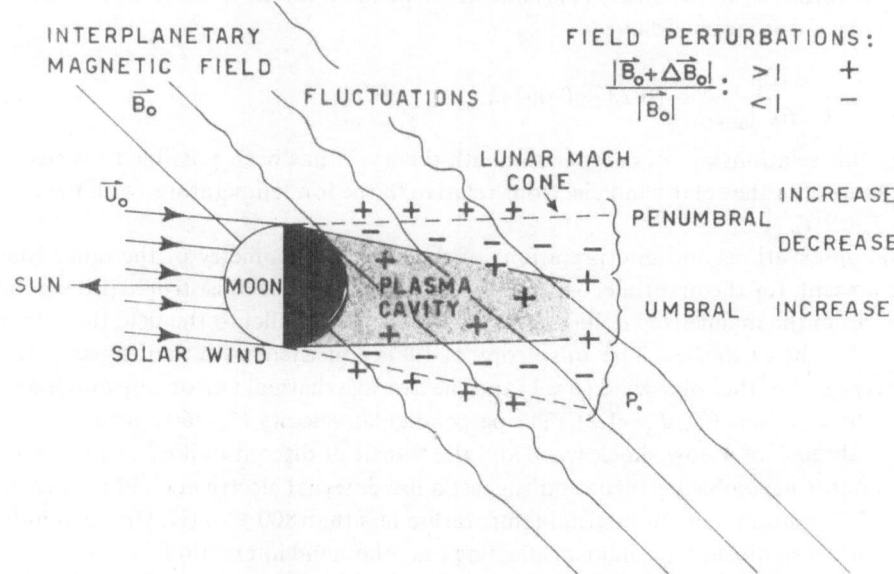

Fig. 24.   Summary view of the 'weak' interaction of the solar wind with the Moon. As the lunar surface absorbs all plasma and particle flux which strike it, a plasma cavity develops and small magnetic field perturbations (<30%) are observed in the plasma umbra and penumbra (characteristically with alternating sign + − + − +).

wake. The Moon appears to behave as a non-magnetic, electrically non-conducting fully absorbing spherical obstacle in the solar wind flow. The principal features of the plasma-field perturbations of the plasma and field (see Figure 24) are:

(1) A downwind *plasma umbral cavity* or void containing an *enhanced interplanetary magnetic field* ($< +30\%$) only slightly perturbed in direction (*no* bow shock wave, pseudo-magnetosphere, or trailing shock in the umbra is observed);

(2) A downwind *penumbral region* aft of a rarefaction wave or Mach cone, elliptical in cross-section, containing a *reduced plasma flux* and *magnetic field* ($> -30\%$);

(3) A very *limited* penumbral region, upwind of the lunar Mach cone, sometimes contains an *enhanced magnetic field* ($< +30\%$) and *plasma flux*; and

(4) A *broad region both upstream and downstream* from the lunar wake, but connected to it by the interplanetary magnetic field, in which *rapid fluctuations* ($\Delta|B|/B_0 \approx$ 0.2) of the magnetic field occur with an amplitude that decreases with distance from the wake.

It is further found that:

(5) The Moon hardly affects the time profile of discontinuity surfaces as they are convectively transported past the Moon (or a shock wave as it propagates past the Moon); and

(6) The *axis of the plasma cavity*, as measured by the lunar Mach cone, is aberrated by $4.5° \pm 0.5°$ from the Moon-Sun line.

The magnitude of the anomaly (1) is linearly dependent on the diamagnetic properties of the plasma as measured by:

$$\left.\frac{\Delta|B|}{B_0}\right|_{\text{umbra}} \doteq (0.23 \pm 0.09)\,\beta_{i+e}. \tag{24}$$

From this relationship, in comparison with theory, it has been possible to derive the temperature of the solar wind electrons relative to the ion temperature, with the result that $T_e \approx 2\,T_i$.

The observations and interpretations of the elliptical geometry of the lunar Mach cone present, for the first time, experimental evidence for the anisotropic propagation of waves in the magnetized collisionless solar plasma. Parallel to the field the velocity is found to be 41 km/sec. The anisotropy of 1.4 is consistent with that expected from the average β of the solar wind ($\beta \approx 1$) and the average thermal anisotropy of the warm magnetized plasma ($T_\parallel/T_\perp \sim 1.2$). (The perpendicular velocity $V_\perp = 60$ km/sec.)

The absence of a bow shock wave and the transit of discontinuities past the Moon with almost negligible perturbation suggest a low internal electrical conductivity, less than 10^{-4} mho/m, and an internal temperature less than 800 ± 200 K. Unipolar induction is not a significant mechanism affecting the solar wind interaction.

This low temperature indicates either that the Moon was formed only recently, if its radioactive composition is similar to chondritic meteorites, or that if it is old (as recent age dating of its surface materials indicates), then its radioactive constituents are much less than those of chondritic meteorites. Urey and MacDonald (1970) have recently reviewed all the relevant data concerning the present physical state of the Moon. They discuss the evidence that there is rough agreement on the maximum temperature that can be derived from the stress differences inside the Moon and from the electrical conductivity. They conclude that the Moon's interior is relatively cold and that there is no extensive region where the temperature exceeds 1000–1200 °C.

5.1. FUTURE PROBLEMS

The data obtained from Explorer 35 and its interpretations have clarified most features of the solar wind interaction with the Moon. However, certain problems remain to be solved, some of which are associated with the fact that Explorer 35 has a periselene greater than 700 km. These problems are:

(1) What is the source mechanism producing the positive penumbral anomalies sometimes observed in the interplanetary magnetic field? (See Subsections 3.1 and 4.1.)

(2) What is the nature of the source cf the stimulated waves propagated away from the lunar wake? (See Subsection 3.3.)

(3) What is the nature of the boundary between solar wind and the lunar surface? Are the effects of finite Larmor radii responsible for the penumbral increases?

(4) How far behind the Moon is its wake detectable and by what means? Does a trailing shock ever exist? (See Subsection 3.2.)

(5) Is Io another Moon-like object in its interaction with the Jovian magnetosphere? (Schatten and Ness, 1970) Or Mercury with the solar wind?

(6) What is the detailed profile of internal electrical conductivity? (See Section 4.)

(7) What is the nature of the detailed plasma flow in the lunar wake region and how are the plasma parameters related to other observed phenomena in the magnetic field?

Acknowledgements

I appreciate the hospitality of the staff of the Consiglio Nazionale delle Ricerche University of Rome Laboratory for Space Plasmas in the preparation of this manuscript, the discussions of the Mach cone phenomena with Dr. Y. C. Whang, and the helpful comments of Dr. F. C. Michel.

References

Behannon, K. W.: 1968, 'Intrinsic Magnetic Properties of the Lunar Body', *J. Geophys. Res.* **73**, 7257

Blank, J. L. and Sill, W. R.: 1969a, 'Response of the Moon to the Time Varying Interplanetary Magnetic Field', *J. Geophys. Res.* **74**, 736.

Blank, J. L. and Sill, W. R.: 1969b, 'Reply', *J. Geophys. Res.* **74**, 5175.

Colburn, D. S., Currie, R. G., Mihalov, J. D., and Sonett, C. P.: 1967, 'Diamagnetic Solar Wind Cavity Discovered behind Moon', *Science*, **158**, 1040.

Debye, P.: 1909, 'Der Lichtdruck auf Kugeln von beliebigen Material', *Ann. Phys.* **30**, 57.

Dessler, A. J.: 1967, 'Ionizing Plasma Flux in the Martian upper Atmosphere', in *Atmospheres of Venus and Mars* (ed. by J. C. Brandt and M. B. McElroy), Gordon and Breach, London, p. 241.

Dolginov, Sh. Sh., Yeroshenko, Ye. G., Zhuzgov, L. N., and Pushkov, N. V.: 'Investigation of the Magnetic Field of the Moon', *Geomagnetizm i Aeronomiya* **1**, 21.

Dolginov, Sh. Sh., Yeroshenko, Ye. G., Zhuzgov, L. N., and Tyurmina, L.O.: 1960, 'Magnetic Measurements with the Second Cosmic Rocket', *Iskusstvennye Sputniki Zemli* **5**, 16.

Dolginov, Sh. Sh., Yeroshenko, Ye. G., Zhuzgov, L. N., and Zhulin, I. A.: 1967, 'Possible Interpretation of the Results of Measurements on the Lunar Orbiter Luna 10', *Geomagnetizm i Aeronomiya* **7**, 436.

England, A. W., Simmons, G., and Strangway, D.: 1968, 'Electrical Conductivity of the Moon', *J. Geophys. Res.* **73**, 3219.

Fahleson, U.: 1967, 'Laboratory Experiments with Plasma Flow Past Unmagnetized Obstacles', *Planetary Space Sci.* **15**, 1489.

Fuller, B. D. and Ward, S. H.: 1969, 'Discussion of the Paper by J. L. Blank and W. R. Sill, "Response of the Moon to the Time Varying Interplanetary Magnetic Field"', *J. Geophys. Res.* **74**, 5173.

Gringauz, K. I., Bezrukikh, V. V., Khokhlov, M. Z., Zastenkev, G. N., Remizov, A. P., and Muzatov, L. S.: 1966, 'Experimental Results Concerning the Lunar Ionosphere Detected on the First Artificial Lunar Satellite', *Dokl. Akad. Nauk. S.S.S.R.* **170**, 1306.

Gold, T.: 1966, 'The Magnetosphere of the Moon' in *The Solar Wind* (ed. by R. I. Mackin, Jr. and M. Neugebauer), Pergamon Press, New York, p. 381.

Greenstadt, E. W.: 1965, 'Interplanetary Magnetic Effects of Solar Flares: Explorer 18 and Pioneer 5', *J. Geophys. Res.* **70**, 5451.

Hirshberg, J.: 1966, 'Discussion of the Paper by N. F. Ness, "The Magnetohydrodynamic Wake of the Moon"', *J. Geophys. Res.* **71**, 4202.

Hollweg, J. V.: 1968, 'Interaction of the Solar Wind with the Moon and Formation of a Lunar Limb Shock Wave', *J. Geophys. Res.* **73**, 7269.

Hollweg, J. V.: 1970, 'Lunar Conducting Islands and the Formation of a Lunar Limb Shock Wave', *J. Geophys. Res.* **75**, 1209.

Ivanov, K. G.: 1965, 'Did the IMP-1 observe the Magnetic Wake of the Moon or Earth?', *Geomagnetizm i aeronomia* **5**, 581.

Johnson, F. and Midgley, J. E.: 1968, 'Notes on the Lunar Magnetosphere', *J. Geophys. Res.* **73** 1523.

Krall, N. A. and Tidman, D. A.: 1969, 'Magnetic Fluctuations near the Moon', *J. Geophys. Res.* **74**, 6439.

Lin, R. P.: 1968, 'Observations of Lunar Shadowing of Energetic Particles', *J. Geophys. Res.* **73**, 3066.

Lüst, V. R.: 1959, 'Uber die Ausbreitung von Wellen in einem Plasma', *Fortschritte der Physik* **7**, 503.

Lyon, E. F., Bridge, H. S., and Binsack, J. H.: 1967, 'Explorer 35 Plasma Measurements in the Vicinity of the Moon', *J. Geophys. Res.* **72**, 6113.

Michel, F. C.: 1964, 'Interaction between the Solar Wind and the Lunar Atmosphere', *Planetary Space Sci.* **12**, 1075.

Michel, F. C.: 1965, 'Detectability of Disturbances in the Solar Wind', *J. Geophys. Res.* **70**, 1.

Michel, F. C.: 1967, 'Shock Wave Trailing the Moon', *J. Geophys. Res.* **72**, 5508.

Michel, F. C.: 1968a, 'Magnetic Field Structure behind the Moon', *J. Geophys. Res.* **73**, 1533.

Michel, F. C.: 1968b, 'Lunar Wake at large Distances', *J. Geophys. Res.* **73**, 7277.

Ness, N. F.: 1965a, 'The Magnetohydrodynamic Wake of the Moon', *J. Geophys. Res.* **70**, 517.

Ness, N. F.: 1965b, 'Remarks on Preceding Paper by E. W. Greenstadt', *J. Geophys. Res.* **70**, 5453.

Ness, N. F.: 1966a 'Reply', *J. Geophys. Res.* **71**, 4205.

Ness, N. F.: 1966b, 'Remarks on the Paper by K. G. Ivanov', *Geomagnetizm i Aeronomiya* **6**, 301.

Ness, N. F.: 1967, 'Remarks on the Interpretation of the Luna 10', *Geomagnetizm i Aeronomiya* **7**, 452.

Ness, N. F.: 1968, Recent Results from Lunar Explorer 35, NASA-GSFC preprint X616-68-335. (To be published in Proceedings of Kiev Conference on the Moon and Planets, October, 1968.)

Ness, N. F.: 1969, 'Lunar Explorer 35', *Space Res.* **9**, 678.

Ness, N. F., Behannon, K. W., Scearce, C. S., and Cantarano, S. C.: 1967, 'Early Results from the Magnetic Field Experiment on Explorer 35', *J. Geophys. Res.* **72**, 5769.

Ness, N. F., Behannon, K. W., Taylor, H. E., and Whang, Y. C.: 1968, 'Perturbations of the Interplanetary Magnetic Field by the Lunar Wake', *J. Geophys. Res.* **73**, 3421.

Ness, N. F., Scearce, C. S. and Seek, J. B.: 1964, 'Initial Results of the IMP-1 Magnetic Field Experiment', *J. Geophys. Res.* **69**, 3531.

Ness, N. F. and Schatten, K. W.: 'Detection of Interplanetary Magnetic Field Fluctuations Stimulated by the Lunar Wake', *J. Geophys. Res.* **74**, 6425.

Neugebauer, M.: 1960, 'Question of the Existence of a Lunar Magnetic Field.' *Phys. Rev. Letters* **4**, 6.

Ogilvie, K. W. and Ness, N. F.: 1969, 'Dependence of the Lunar Wake on Solar Wind Plasma Characteristics', *J. Geophys. Res.* **74**, 4123.

Schatten, K. W. and Ness, N. F.: 1970, 'Modulation of Jupiter's Radio Emission by Io', *Trans. AGU* **52**, 481.

Schneider, O.: 1967, 'Interaction of the Moon with the Earth's Magnetosphere', *Space Sci. Rev.* **6**, 655.

Schubert, G. and Schwartz, K.: 1969, 'A Theory for the Interpretation of Lunar Surface Magnetometer Data', *The Moon* **1**, 106.

Schwartz, K. and Schubert G.: 1969, 'Time-Dependent Lunar Electric and Magnetic Fields Induced by a Spatially Varying Interplanetary Magnetic Field', *J. Geophys. Res.* **74**, 4777.

Schwartz, K., Sonett, C. P., and Colburn, D. S.: 1969, 'Unipolar Induction in the Moon and a Lunar Limb Shock Mechanism', *The Moon* **1**, 7.

Sill, W. R. and Blank, J. L.: 1970, 'Method for Estimating the Electrical Conductivity of the Lunar Interior', *J. Geophys. Res.* **75**, 201.

Siscoe, G. L., Lyon, E. F., Binsack J. H., and Bridge, H. S.: 1969, 'Experimental Evidence for a Detached Lunar Compression Wave', *J. Geophys. Res.* **74**, 59.

Sonett, C. P. and Colburn, D. S.:1967, 'Establishment of a Lunar Unipolar Generator and Associated Shock and Wake by the Solar Wind', *Nature* **216**, 340.
Sonett, C. P. and Colburn, D. S.: 1968, 'The Principle of Solar Wind Induced Planetary Dynamos', *Phys. Earth Planetary Int*. **1**, 326.
Sonett, C. P., Colburn, D. S., and Currie R. G.: 1967, 'The Intrinsic Magnetic Field of the Moon', *J. Geophys. Res*. **72**, 5503.
Spreiter, J. and Alksne, A.:1969, 'Plasma Flow around the Magnetosphere, *Rev. Geophys*. **7**, 11.
Taylor, H. E.: 1968, Aspect Determination in Lunar Shadow on Explorer 35, NASA-TN D-4544.
Taylor, H. E., Behannon, K. W., and Ness, N. F.: 1968, 'Measurements of the Perturbed Interplanetary Magnetic Field in the Lunar Wake', *J. Geophys. Res*. **73**, 6723.
Tozer, D. C. and Wilson, J., III: 1967, 'The Electrical Conductivity of the Moon, The Electrical Conductivity of the Moon's Interior', *Proc. Roy. Soc. London Ser. A*, **296**, 320.
Urey, H. C. and MacDonald, G. J. F.: 1970, 'Origin and History of the Moon', revised Chapter 13 in *Physics and Chemistry of the Moon* (ed. by Z. Kopal), Academic Press, New York.
Van Allen, J. A. and Ness, N. F.:1969, 'Particle Shadowing by the Moon', *J. Geophys. Res*. **74**, 71.
Wait, J. R.: 1951, 'A Conducting Sphere in a Time Varying Field,' *Geophys*. **16**, 666.
Whang, Y. C.: 1968a, 'Interaction of the Magnetized Solar Wind with the Moon', *Phys. Fluids* **11**, 969.
Whang, Y. C.: 1968b, 'Theoretical Study of the Magnetic Field in the Lunar Wake', *Phys. Fluids* **11**, 1713.
Whang, Y. C.: 1969, 'Field and Plasma in/the Lunar Wake', *Phys. Rev*. **186**, 143.
Whang, Y. C. and Ness, N. F.: 1970, Observations and Interpretations of the Lunar Mach Cone, NASA-GSFC preprint X-692-70-60.
Wolf, R. A.: 1968, 'Solar Wind Flow behind the Moon', *J. Geophys. Res*. **73**, 4281.
Zhuzgov, L. N., Dolginov, Sh. Sh., and Yeroshenko, Ye. G.: 1966, 'Investigation of the Magnetic Field from the Satellite Luna 10', *Kosmicheskiye Issledovaniya* **4**, 880.